10/18/89

D1283815

NON-METAL RINGS, CAGES AND CLUSTERS

NON-METAL RINGS, CAGES AND CLUSTERS

J. D. Woollins
Department of Chemistry
Imperial College of
Science and Technology,
South Kensington,
London SW7 2AY

John Wiley & Sons

Chichester · New York · Brisbane · Toronto · Singapore

Library of Congress Cataloging in Publication Data:

Woollins, J. D. (J. Derek)
  Non-metal rings, cages and clusters.

  Includes bibliographies and index.
  1. Nonmetals 2. Cyclic compounds. I. Title.
QD161.W66 1988   546'.7   88–5570
ISBN 0 471 91592 0

British Library Cataloguing in Publication Data:

Woollins, J. D. (J. Derek)
  Non-metal rings, cages and clusters.
  1. Non metals. Structure & chemical
  properties
  I. Title
  546'.7
ISBN 0 471 91592 0

Phototypesetting by Thomson Press (India) Limited, New Delhi
Printed in Great Britain by Anchor Brendon Ltd., Essex

*In memory of my dad
who recognized the true worth
of an education*

CONTENTS

PREFACE

There are some excellent texts which deal with particular areas of the chemistry of non-metal rings, cages and clusters in great depth but, with one or two exceptions (e.g. boron hydride chemistry) there are no useful introductions to this topic. This observation was brought home to me when, after one of my (typically rambling) undergraduate lectures, a student asked me to recommend a good book. Unfortunately, I could not do so. I have no misconceptions that this book will totally solve the problem. My main goal is to introduce the area to a wide audience which appears to be currently almost unaware of the existence of many of the compounds, or has difficulty because 'there are so many different shapes and structures.' This book is arranged along lines intended to accentuate the structural similarities in a rather diverse area. In order to keep the book short and (hopefully) coherent, I have mercilessly chosen to ignore great tracts of chemistry (I do regard silicates as important!). In a way the coverage should be regarded as my choice from a very large box of chocolates—I prefer the hard centres and so have gone for them first, but this does not mean that my choice is exclusive or objective. I make no apologies for this. If this book succeeds in interesting students in non-metal or cluster chemistry then it will not have been in vain.

Many people have helped, directly or indirectly, during the writing of this book. The work was carried out at Imperial College of Science and Technology and I am grateful to Prof. Sir G. Wilkinson FRS for his guidance, support and friendship. My research group have soldiered on in my absence and, perhaps more to the point, in my bad-tempered presence. It is a great pleasure to acknowledge my wife, Ann, for proof reading and editing various drafts. Finally, I am grateful to John Gray for his excellent efforts in converting my scribbles into diagrams.

December 1987   Derek Woollins

CHAPTER ONE

Introduction

1.1 INTRODUCTION

The majority of inorganic textbooks contain passing references to main group ring and cluster compounds, but rarely make any effort to consider them as a class. There are significant problems in attempting to group together all of the main group rings, heterocycles and clusters. However, there are also enormous benefits, especially if the apparently endless range of species is logically organized by virtue of the structural similarities, although at first glance many of the compounds appear to be in unrelated branches of chemistry. Unfortunately, at present, there is no simple all encompassing theoretical treatment which allows us to understand and relate together the known species. The range of compounds includes boron hydride clusters in which sophisticated bonding descriptions are needed, phosphazenes, which can be usefully compared with organic aromatics, anions and cations such as Sn_9^{4-} and Se_4^{2+}, and simple binary phosphorus sulphides such as P_4S_3. Clearly, we cannot hope to understand all of these compounds with one theoretical treatment. Indeed, many chemists would doubt the virtue of starting from a theoretical standpoint—practical observations are the basis of the subject. However, we should attempt to arrange and classify observations into patterns; not only does it improve our understanding, it also reduces the apparent complexity of the subject.

I have chosen to divide this book into sections in which the compounds are loosely defined as 'electron-deficient,' 'electron-precise/classical' or 'electron-rich.' This method of arrangement is indicative of my preconceptions and bias; other authors might have preferred to separate the material into Groups of the Periodic Table, but this has problems. For example, borazenes would be in the same chapter as the boron–hydrides, which have totally different properties.

The classification of compounds as electron-deficient, etc., is based on the excellent review by Gillespie[1] and is outlined below. In this introductory chapter, I also provide a brief outline of reaction types employed in the formation of inorganic rings and clusters.

This book provides only an introduction to the subject. There are a number of excellent monographs available and these are mentioned as appropriate. Probably the most complete descriptions are in the various supplements to

Gmelins Handbook of Inorganic Chemistry, but this is an extremely expensive series and not all libraries subscribe to it.

Throughout the text I have tried to provide up-to-date references from the major chemical journals. However, I do not claim to have made any attempt at complete literature surveys—introductory references and reviews are given wherever possible.

The naming of inorganic rings and clusters is difficult and often rather unwieldy. In general I have used the 'trivial' common names by which the various species are most well known. Unfortunately, this is an area in which systematization is desperately needed. Often rings are named after similar organic molecules whilst, with the exception of boranes, naming many clusters is almost impossible!

1.2 FUNDAMENTALS—DEFINING A CLUSTER BY COUNTING ELECTRONS

An electron-precise cluster obeys the following rules: (1) simple localized bonding treatment are adequate; (2) each atom in the cluster obeys the octet rule; (3) each vertex of the cluster is three co-ordinate or, put another way, has a connectivity of three; and (4) each cluster atom has two electrons in a lone pair or a bond to an exocyclic group (Fig. 1.1). Combining (1)–(4) it automatically follows that an electron-precise n-atom cluster will have $5n$ electrons of which $3n$ are involved in cluster bonding.

To calculate the number of electrons in a cluster, total all of the electrons coming from the cluster vertex atoms **plus** any electrons in the vertex atom valence shells from exocyclic atoms, e.g. H atoms (thus S, NH and P^- all have 6e, whereas P, CH and CR have 5e). For example:

P_4 Electron configuration of $P = s^2p^3 = 5$ valence electrons. Total electron count $4 \times 5 = 20$ electrons $= 5n$, i.e. electron-precise

$(C^tBu)_4$ Valence shell of each cluster carbon atom contains $s^2p^2 = 4$ electrons from cluster carbon plus $1\,e^-$ from exocyclic ^tBu group, i.e. electron-precise (in C—C bond). Total electron count 4×5 electrons, i.e. electron-precise.

P_4 is isostructural with $(C^tBu)_4$.

S_8 is isoelectronic with $S_7(NH)$ and $S_4(NH)_4$. Total electron counts are $8 \times 6 = 48$, i.e., greater than $5n = 40$, electron-rich and therefore open ring structures are observed.

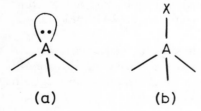

(a)   **(b)**

Figure 1.1 (a) A three-connected atom A as observed in P_4; (b) non-bonding lone pair relaced by exocyclic atom (or group), e.g. H, Cl, R.

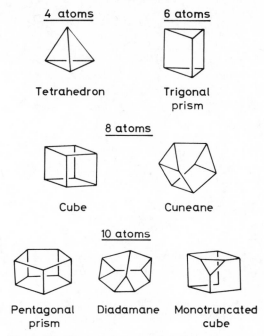

4 atoms

Tetrahedron

6 atoms

Trigonal prism

8 atoms

Cube

Cuneane

10 atoms

Pentagonal prism

Diadamane

Monotruncated cube

Figure 1.2 Cluster frameworks for electron-precise 4-, 6-, 8- and 10-vertex clusters.

Figure 1.2 shows the cluster frameworks for 4-, 6-, 8- and 10-atom electron precise clusters for which there are known examples. Many of these examples come from organic chemistry, e.g. tetrahedrane and prismane.

A cage is electron-rich, has greater than $5n$ electrons, contains some 2-coordinate atoms, and in consequence has a more open structure than a cluster. In rings all of the atoms are 2-coordinate and the total number of electrons is $6n$ (e.g. S_6, S_8). However, the number of electrons actually involved in bonding can vary [e.g. S_6 vs $(PNCl_2)_3$—comparable to cyclohexane vs benzene in organic chemistry]. In electron-deficient clusters with less than $5n$ electrons (e.g. boranes) the first rule is not adhered to and three-centre, two-electron bonding descriptions are needed. The bonding in electron-deficient clusters is discussed on p. 12; for the moment we shall limit ourselves to describing the scheme used to classify electron-precise and electron-rich systems.

Figure 1.3 plots the number of cluster/ring bonding electrons against the position of the cluster atoms in the Periodic Table. As might be expected, those species containing Group III atoms are electron-deficient, those with Group V atoms are electron-precise and those with Group VI atoms are electron-rich. An important feature of virtually all of the examples in Fig. 1.3 is that on going from left to right, i.e. increasing the number of cluster bonding electrons, more open structures result, a concept which was pointed out many years ago.[2]

4

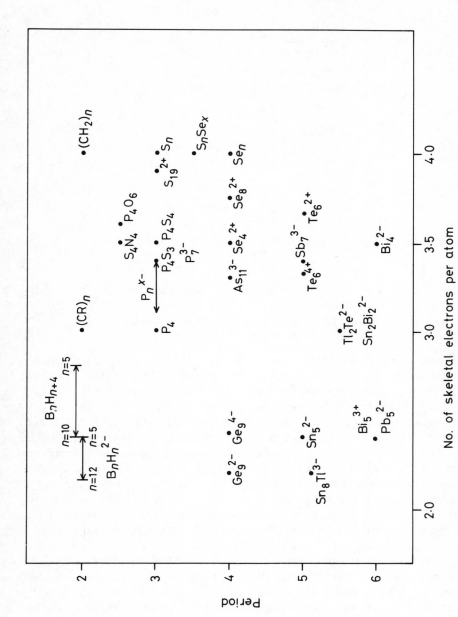

Figure 1.3 Number of cluster bonding electrons per atom versus position in the Periodic Table.

Gillsepie[1] has illustrated how a wide range of cages and rings are related to the basic cluster frameworks in Fig. 1.2. Figure 1.4 summarizes the transformations which can be used to accommodate the extra electrons into cages and rings: (1) one or more edges of a cluster may be bridged; (2) one or more edges of a cluster may be broken by the addition of an electron pair (the edge bond is replaced by two lone pairs); (3) one or more atoms may be removed leaving the bonding electrons as lone pairs; and (4) removal of a 2-coordinate bridge atom from an edge has the same overall effect as (2).

The above rules can be used to generate most of the structures in Chapters 3 and 4 *without the need for a detailed understanding of the bonding*. The major use of

Edge bridge

E.g. bridging three edges of a P_4 tetrahedron with S atoms or P^- gives P_4S_3 and P_7^{3-}, respectively

(P_4S_3 structure, Section 3 18)

or Bridging S_4N_4 with N^-

E.g. removal of S^{2+} from P_4S_3 or P^+ from P_7^{3-} gives $Te_3Se_3^{2+}$

E = P^- or S

Lone pair addition

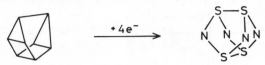

E.g. adding $4e^-$ to the cuneane structure gives the S_4N_4 or P_4S_4 cage

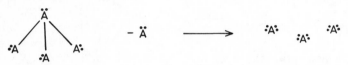

Atom removal

E.g. removal of two atoms from a single bridged cube, with one edge broken, gives the S_7 structure

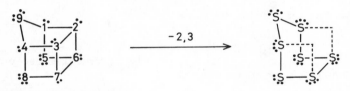

Removal of bridge atom

Figure 1.4 Rules and examples for deriving cage structures.

this type of description is to enable us to relate different species and to order our thinking. However, it should be noted that it is a very simple empirical treatment and as such may fail in some circumstances.

The most noticeable features of the transformations in Fig. 1.4 is that more open structures are associated with electron-rich species. The excess electrons are generally accommodated in additional lone pairs, i.e. some atoms have two lone pairs rather than one. Further examples of this effect are shown in Fig. 1.5.

1.3 SYNTHETIC STRATEGIES

Although there is a very wide range of cyclic/cluster compounds, it is possible to divide synthetic routes into a few general classes. The types or reactions are indicated below.

(1) *Combination of the elements* is most generally applicable to phosphorus sulphides, e.g.

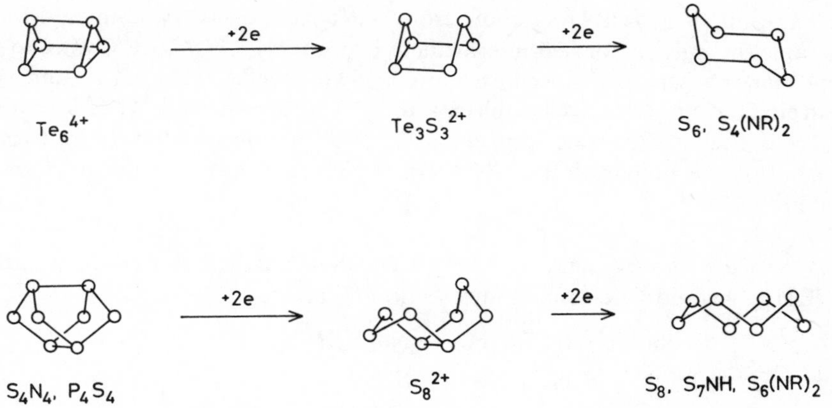

Te_6^{4+} $+2e$ $Te_3S_3^{2+}$ $+2e$ S_6, $S_4(NR)_2$

S_4N_4, P_4S_4 $+2e$ S_8^{2+} $+2e$ S_8, S_7NH, $S_6(NR)_2$

Figure 1.5 Effect of addition of electron pairs to clusters.

$$P_4 + 10/8S_8 \longrightarrow P_4S_{10}$$

(2) *Polymerization* is useful in a few cases, although reactions can be complex.

$$3NSF \longrightarrow (NSF)_3$$

$$B_2H_6 \xrightarrow{\Delta} B_5H_9 + B_6H_{10} + B_{10}H_{14}$$

$$P_2H_4 \longrightarrow P_3H_5$$

(3) *Cyclocondensation/cycloaddition reactions* have wide application. Usually the driving force for these reactions is the elimination of a stable non-cyclic species, either a salt [e.g. in the preparation of $S_4(NR)_2$] or the formation of a very strong bond [e.g. Sn—F in the preparation of S_2N_2CO].

$$P_4S_3I_2 + (Me_3Sn)_2S \longrightarrow P_4S_4 + 2Me_3SnI$$
$$1,3\text{-}S_6(NH)_2 + S_5Cl_2 \longrightarrow S_{11}N_2 + 2HCl$$
$$Me_2SnS_2N_2 + COF_2 \longrightarrow S_2N_2CO + Me_2SnF_2$$
$$Me_2SiCl_2 + H_2O \longrightarrow 1/n(Me_2SiO)_n + 2HCl$$
$$PCl_5 + NH_4Cl \longrightarrow 1/n(NPCl_2)_n + 4HCl$$
$$2S_2Cl_2 + 6EtNH_2 \longrightarrow S_4(NEt)_2 + 4[NH_3Et]Cl$$
$$Cp_2TiS_5 + SeCl_2 \longrightarrow SeS_5 + Cp_2TiCl_2$$
$$RPCl_2 + RHPPHR \longrightarrow (PR)_3 + 2HCl$$

In the above examples, the products obtained are the result of kinetic or thermodynamic control. High dilution conditions (Ruggli–Ziegler) which involve continuous slow mixing of the two reactants are very useful, e.g. in the preparation of $1,4\text{-}(RN)_2S_4$ rings from S_2Cl_2 and primary amines. In this example the initially formed intermediate is probably ClS_2NHR. Owing to the conditions chosen, this molecule is more likely to meet another of the same type than either an amine or S_2Cl_2 molecule, thus favouring cyclization and formation of the six-membered ring. An example of a thermodynamically controlled reaction is the preparation of $(NPCl_2)_n$, which is carried out under vigorous conditions.

(4) *Ring expansion reactions* are very useful. Some of these (e.g. the preparation of $B_6H_{11}^-$) could be classified into section (3) above.

$$NaB_5H_8 + 1/2B_2H_6 \longrightarrow Na[B_6H_{11}]$$

$$S_4N_4 + NS^+ \longrightarrow S_5N_5^+$$

$$P_4 + \text{excess } LiPH_2 \longrightarrow Li_3P_7$$

(5) *Ring contraction/cluster breakdown reactions* are widespread, but not always predictable!

$$B_{10}H_{14} + KOH \longrightarrow K[B_9H_{14}]$$

$$S_4N_4 \xrightarrow[\text{Ag wool}]{\Delta} S_2N_2$$

$$P_4S_{10} + 3PPh_3 \longrightarrow P_4S_7 + 3SPPh_3$$

$$P_4S_{10} + MeOPh \longrightarrow MeOC_6H_4P(S)S_2P(S)C_6H_4OMe$$

$$(PhP)_5 + Se_8 \longrightarrow Ph(Se)PSe_2P(Se)Ph$$

Apart from the above groups of reactions, there are also 'simple' oxidation/reduction reactions such as the formation of S_8^{2+} from S_8 or the preparation of $S_4N_4H_4$ from S_4N_4. These reactions do not involve the formation of a new ring, although they may result in quite large structural changes.

1.4 REFERENCES

1. R. J. Gillespie, *Chem. Soc. Rev.*, 1979, **8**, 315.
2. D. M. P. Mingos, *Nature Phys. Sci.*, 1972, **236**, 99.

An excellent, comprehensive, survey of main group rings and cages has recently been published. '*The Chemistry of Inorganic Homo- and Heterocycles*', Vols. 1 and 2, 1987, I. Haiduc and D. B. Sowerby, Academic Press, London.

CHAPTER TWO

Electron-Deficient Species

2.1 BORANES

2.1.1 Introduction—Nomenclature

After hydrocarbons, boranes (boron hydrides) are the largest class of hydrides, stretching from BH_4^- to $B_{20}H_{16}$ with a wide variety of known structures (Fig. 2.1). We shall consider them in some detail since they are illustrative of cluster bonding theories—an area in which they have played an important part. Although there are very many isoelectronic carboranes (C≡BH), because of space limitations we shall not make more than passing reference to their chemistry.

Naming boranes is non-trival and there have been some suggestions made recently with regard to nonmenclature;[1] we shall limit ourselves to the most important aspects which are well accepted. For neutral molecules the number of boron atoms is indicated by a prefix with the number of hydrogen atoms being shown at the end of the name. For example, B_2H_6 is named diborane(6) and B_6H_{10} is hexaborane(10). The names of anionic species are terminated by *ate* with the charge being indicated in parentheses, e.g. $B_5H_8^-$ is called octahydropentaborate(-1). Additionally, prefixes which identify the general formulae are usually employed. Illustrative examples are shown below; we shall discuss this aspect again when we consider the bonding descriptions.

Prefix	General formula	Type of structure
closo	$B_nH_n^{2-}$	'closed'—clusters with only terminal BH bonds
nido	B_nH_{n+4}	'nest' structures with BHB bridges
arachno	B_nH_{n+6}	'web'—more open structures
hypho	B_nH_{n+8}	'net'—most open clusters
conjuncto	—	fused or linked combinations of the above boranes

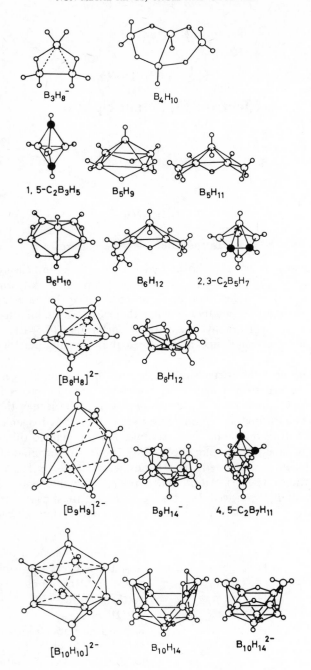

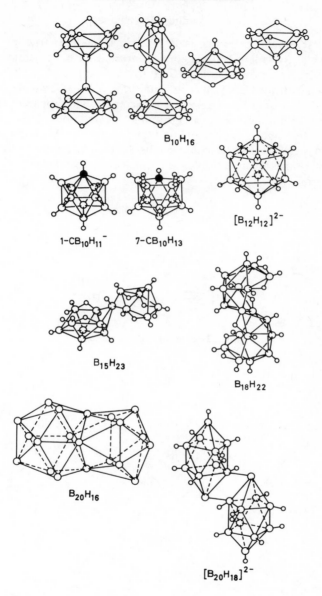

$B_{10}H_{16}$

$1-CB_{10}H_{11}^-$ $7-CB_{10}H_{13}$

$[B_{12}H_{12}]^{2-}$

$B_{15}H_{23}$

$B_{18}H_{22}$

$B_{20}H_{16}$

$[B_{20}H_{18}]^{2-}$

Figure 2.1 Structures of boranes. Carborane examples are shown where the parent borane is either unknown or less well characterized.

2.1.2 Bonding Descriptions

The simplest isolatable borane is B_2H_6, which has the structure shown in Fig. 2.2 with effectively tetrahedral geometry around the boron atoms. The total number of valence electrons available for bonding in diborane is 12 (each boron has 3 and each hydrogen has 1). If each connection in Fig. 2.2 were a simple two-electron bond, we would require $8 \times 2 = 16$ electrons. This apparent shortage of electrons leads to boranes being categorized as 'electron-deficient.' Clearly, in a classical sense this label is satisfactory and the arrangement of this book is based on a similar type of classification, **but** it must be remembered that the chemical properties of boranes are not always those which are associated with molecules in need of electrons.

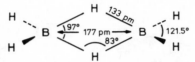

Figure 2.2 Structure of diborane(6).

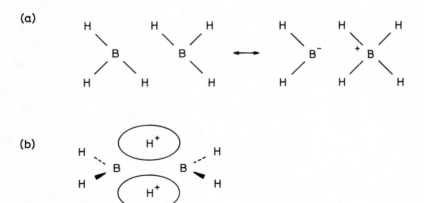

Figure 2.3 (a) Valence bond and (b) protonated double bond descriptions of the bonding in diborane(6).

A number of approaches have been used to attempt to rationalize the structure of diborane, for example valence bond or protonated double bond treatments (Fig. 2.3). The most useful description (from Longuet-Higgins) involves three-centre, two-electron BHB bonds. It is assumed that the boron atoms are sp^3 hybridized and form simple two-centre, two-electron bonds to their terminal hydrogens. The two remaining sp^3 hybrids of each boron and the hydrogen 1s orbitals of the bridging hydrogens can be combined as shown in Fig. 2.4 to produce an MO diagram which consists of one bonding and two antibonding orbitals. The three-centre, two-electron (3c–2e) bond is widely accepted and is not limited to BHB systems; BBB units in higher boranes may also be bonded in this way. With these latter systems both 'open' bonds like that in Fig. 2.4 and 'closed' bonds, shown in Fig. 2.5, are possible.

In larger boranes, localized bonding descriptions based on 2c–2e and 3c–2e bonds have been extensively developed by Lipscomb.[2] Individual boranes are

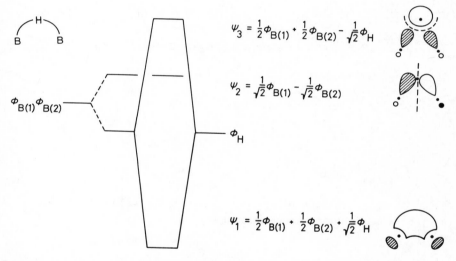

$$\psi_3 = \tfrac{1}{2}\phi_{B(1)} + \tfrac{1}{2}\phi_{B(2)} - \tfrac{1}{\sqrt{2}}\phi_H$$

$$\psi_2 = \tfrac{1}{\sqrt{2}}\phi_{B(1)} - \tfrac{1}{\sqrt{2}}\phi_{B(2)}$$

$$\psi_1 = \tfrac{1}{2}\phi_{B(1)} + \tfrac{1}{2}\phi_{B(2)} + \tfrac{1}{\sqrt{2}}\phi_H$$

Figure 2.4 MO scheme for the ('open') three-centre, two-electron BHB bond in diborane(6).

Figure 2.5 The 'closed' three-centre, two-electron BBB bond.

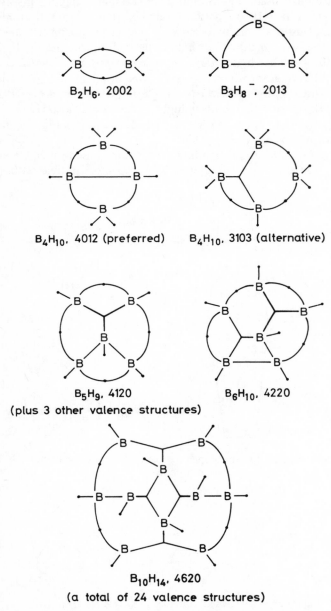

Figure 2.6 Some examples of *styx* topologies.

defined by the so called *styx* topology. Some examples are shown in Fig. 2.6.

s = No. of BHB bridges/3-centre bonds;

t = No. of BBB bridges/3-centre bonds;

y = No. of BB 2c-2e bonds;

x = No. of BH_2 groups.

When attempting to draw localized bonding descriptions, a number of simple rules apply, as follows.

1. Every pair of boron atoms which are geometric neighbours must be connected by a BB, BHB or BBB bond.
2. Every boron atom must use four orbitals and achieve its octet.
3. No two boron atoms may be bonded together by both two-centre and three-centre bonds.
4. The total number of three-centre bonds $(s + t)$ equals the number of boron atoms.
5. $2(s + t + y + x)$ = No. of atoms in a neutral borane.
6. Boranes usually adopt structures with a two-fold element of symmetry.

By using the above rules it is possible to draw simple bonding schemes for a variety of boranes and, with care, structures can also be predicted. However, larger clusters and *closo* compounds do pose problems, particularly the large number of resonance forms that are possible. For example, $B_{10}H_{14}$ can be represented by 24 equivalent valence structures. An alternative approach (which can also be applied to other clusters) is the polyhedral electron skeletal pair theory,[3,4] nowadays often referred to as 'Wade's rules.' The geometries of all the boranes are related to a 'basis set' of n-vertex icosahedral *closo*-boranes whose geometries are minimum-energy systems. Note that these structures all contain triangulated faces.

Each boron provides three orbitals and two electrons for cluster bonding, with additional electrons coming from the charge or the excess hydrogens over B_nH_n. The number of electrons available for cluster bonding is most easily calculated by summing all of the valence electrons and then subtracting $2e^-$ per boron for terminal BH bonding. For example, $B_6H_6^{2-}$ has $6 \times 3e^-$ from boron + 6×1 from the hydrogens + 2 from the negative charge. The total number of e^- is 26. Since there are 6 borons, we subtract $12e^-$, leaving $14e^-$ or 7 electron pairs. For *closo*-boranes of formula $B_nH_n^{2-}$, $n + 1$ electrons are used in cluster bonding; an MO description illustrating why this should be is shown in the next section.

The structures of *nido* and *arachno* clusters are derived from the *closo* compounds by use of the principle that *the number of cluster bonding electrons defines the parent icosahedra*. The structures of B_5H_9 and B_4H_{10} are illustrative examples. B_5H_9 has $(5 \times 3) + (9 \times 1) - (5 \times 2) = 14$ cluster bonding electrons, as

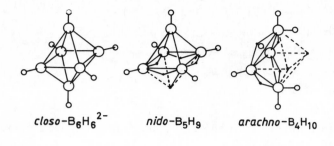

$closo\text{-}B_6H_6{}^{2-}$　　　$nido\text{-}B_5H_9$　　　$arachno\text{-}B_4H_{10}$

iso-arachno structure

Figure 2.7 Structures of boranes which have seven skeletal cluster electron pairs and the related *iso-arachno* geometry.

does B_4H_{10} $[(4 \times 3) + (10 \times 1) - (4 \times 2)]$. Hence we can see that $B_6H_6{}^{2-}$, B_5H_9 and B_4H_{10} have the same number of electrons available for cluster bonding and the structures are all based on an octahedron (Fig. 2.7). These conclusions can be summarized as follows:

closo	$B_nH_n{}^{2-}$	$n + 1\,e^-$ pairs	based on n vertex polyhedra.
nido	B_nH_{n+4}	$n + 2\,e^-$ pairs	based on $n + 1$ vertex polyhedra with 1 vertex unfilled.
arachno	B_nH_{n+6}	$n + 3\,e^-$ pairs	based on $n + 2$ vertex polyhedra with 2 vertices unfilled.

Obviously there are some rather empirical features to this treatment, in particular the choice of which vertex should be left unfilled in *nido* and *arachno* clusters. Normally with *nido* clusters the vertex with the highest connectivity is removed whilst with *arachno* compounds neighbouring vertices are generally removed from the parent polyhedron.

N.B. It is interesting to note that the Libscomb localized bond treatment utilizes the same number of electrons for cluster bonding as the Wade rules method—compare the structures in Fig. 2.6 with our electron counts above.

In carboranes carbon provides a similar number of electrons to the BH group. In general the carbon atoms are situated at the vertices with the lowest

Table 2.1 Classification of the known boron hydrides and related 'naked' clusters

No. of vertices	geometry of base polyhedron	closo $B_nH_n^{2-}$	nido B_nH_{n+2}	arachno B_nH_{n+4}
5	Trigonal bipyramid	$C_2B_3H_5$ Sn_5^{2-}, Pb_5^{2-} Bi_5^{3+}	P_4, $Pb_2Sb_2^{2-}$, Si_4^{2-}, $Sn_4^{2-}*$ iso-nido- $Tl_2Te_2^{2-}$	$B_3H_8^-$
6	Octahedron	$B_6H_6^{2-}$ CB_5H_7 $C_2B_4H_6$	B_5H_9 $C_2B_3H_7$	B_4H_{10} Si_4^{6-} iso-arachno- Sb_4^{2-}, Bi_4^{2-}, $As_4^{2-}*$
7	Pentagonal bipyramid	$B_7H_7^{2-}$ $C_2B_5H_7$	B_6H_{10} $C_xB_{6-x}H_{10-x}$	B_5H_{11}
8	Dodecahedron (D_{2d})	$B_8H_8^{2-}$ $C_2B_6H_8$	—	B_6H_{12}
9	Tricapped trigonal prism	$B_9H_9^{2-}$ $C_2B_7H_9$ Ge_9^{2-}, $TlSn_8^{3-}$, Sn_9^{3-}	B_8H_{12} $C_2B_6H_{10}$	
10	Bicapped square antiprism	$B_{10}H_{10}^{2-}$ $C_2B_8H_{10}$ $TlSn_9^{3-}$	$B_9H_{12}^-$ $C_2B_7H_{11}$ Sn_9^{4-}, Ge_9^{4-} Sn_8Sb^{3-} $TlSn_8^{5-}$ $Sn_xPb_{9-x}^{4-}*$ $Bi_9^{5+\dagger}$	B_8H_{14} Bi_8^{2+} (square antiprism)
11	Octadecahedron (C_{2v})	$B_{11}H_{11}^{2-}$ $C_2B_9H_{11}$	$B_{10}H_{14}$ $C_2B_8H_{12}$	$B_9H_{14}^-$ $C_2B_7H_{13}$
12	Icosahedron	$B_{12}H_{12}^{2-}$ $CB_{11}H_{12}^-$ $C_2B_{10}H_{12}$	$CB_{10}H_{13}^-$ $C_2B_9H_{11}$	$B_{10}H_{14}^{2-}$ $B_{10}H_{12}L_2$

*Observed in solution only.
†Exists as tricapped trigonal prism; see text for discussion.

coordination number and carboranes often rearrange to systems which minimize the number of C–C and maximize the number of C–B connections.

This system is readily extended to other clusters, such as Zintyl anions and transition metal carbonyls, discussed on pp. 32–39. Table 2.1 summarizes the known main-group structures.

Wade's rules provide a straightforward way of rationalizing and, to some extent, predicting structures, but we have not yet developed any picture of what types of orbitals the electrons occupy. A simplified MO description for $B_6H_6^{2-}$ is shown in Fig. 2.8. Starting with sp hybridized boron atoms, the six outward-

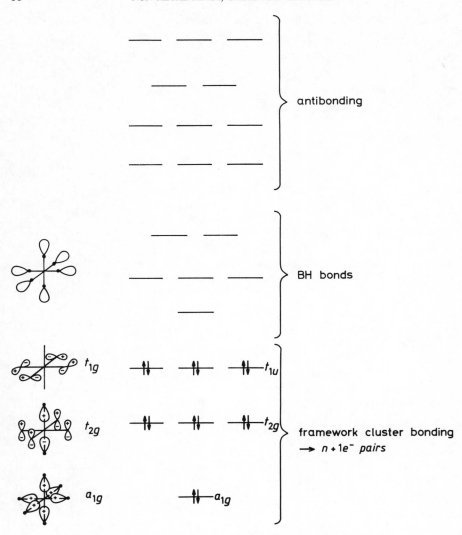

Figure 2.8 Simple MO scheme for the *closo*-$B_6H_6^{2-}$ anion.

pointing sp hybrids combine with hydrogen 1s orbitals to form the BH MOs. The six inward-pointing sp hybrids combine in a unique orbital and in a set of three degenerate orbitals by combination with tangential p orbitals. A final set of bonding orbitals is obtained by combining p orbitals. The most important feature which arises is that there are $n + 1$ cluster bonding orbitals. For a review of this topic, see ref. 5.

2.1.3 Preparative Routes

Much of the original work was carried out by Alfred Stock between 1912 and 1936 and his early book[6] still makes interesting reading. Stock and co-workers isolated and studied B_2H_6, B_4H_{10}, B_5H_9, B_6H_{10} and $B_{10}H_{14}$. This was a remarkable achievement since they also had to develop many of the vacuum techniques needed to handle volatile air-sensitive compounds.

The earliest reactions used magnesium boride and concentrated HCl to give a mixture of boranes in 4–5% combined yield. Later a high-yield (!) route was devised:

$$Mg_3B_2 + H_3PO_4 \longrightarrow B_4H_{10} + B_5H_9 + B_6H_{10}$$
combined yield *ca* 10%

Immediately after World War II the American armed forces sponsored large research programmes, partly in the hope of developing rocket propellants based on boranes. This led to rapid developments in this area. A number of high-yield routes to diborane were developed, and the discovery of borohydride provided a convenient *in situ* synthesis which has found many applications in organic synthesis.[7]

$$2BF_3 + 6NaH \longrightarrow B_2H_6 + 6NaF$$
$$3LiAlH_4 + 4BCl_3 \longrightarrow 2B_2H_6 + 3LiCl + 3AlCl_3$$
$$2NaBH_4 + I_2 \longrightarrow B_2H_6 + 2NaI + H_2$$

The preparation of higher boranes is more difficult[8] and for many years relied on pyrolysis reactions of diborane(6), making use of the hot-tube technique. In this method the reactor consists of hot and cold regions, thus allowing volatile starting material and intermediates to come into contact with a hot surface whilst the cold surface is available for condensation of the products. Yields of $B_{10}H_{14}$ in excess of 50% have been obtained when B_2H_6 is injected via a multijet system on to a block at 320 °C with the walls of the vessel at 0–5 °C. As might be expected, these techniques are very difficult to optimize; for example, lowering the temperature of the flask walls to -40 °C results in B_5H_9 becoming the major product. The mechanism of pyrolysis reactions is only partially understood and is still under investigation.[9]

Reductive disproportionation of diborane(6) is a very useful method for the preparation of higher boranes:

$$B_2H_6 + Na(BH_4) \longrightarrow Na(B_3H_8) + H_2$$

A single-stage method which combines the *in situ* preparation of diborane with the above reaction is particularly convenient. The reaction is carried out at 95 °C in diglyme:

$$3Na(BH_4) + I_2 \longrightarrow 2NaI + 2H_2 + Na(B_3H_8)$$

20

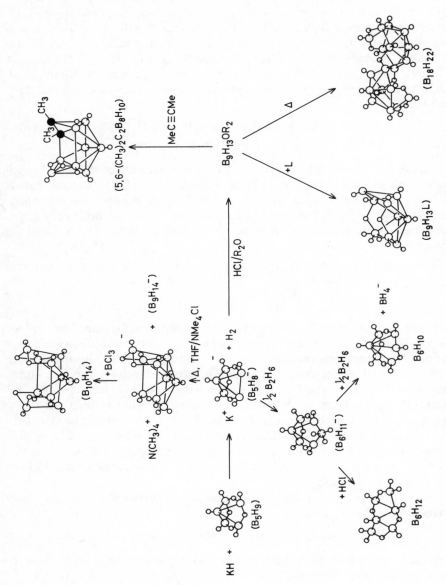

Figure 2.9 Some rational syntheses of boranes and carboranes.

closo-Boranes can be prepared by base-promoted condensations, pyrolytic condensations and degradations of larger *closo* clusters:

$$Et_3N + B_{10}H_{14} \xrightarrow[-H_2]{} arachno\text{-}B_{10}H_{12}(Et_3N)_2 \xrightarrow{heat} closo\text{-}[Et_3NH]_2[B_{10}H_{10}]$$

$$2NaBH_4 + B_{10}H_{14} \longrightarrow Na_2[B_{12}H_{12}]$$

$$Cs_2[B_{10}H_{10}] \xrightarrow{heat} Cs_2[B_9H_9]$$

Some elegant work on rational syntheses of *nido* and *arachno* compounds has been reported[10] and examples are shown in Fig. 2.9. The first stage in many of the preparations is a deprotonation (using KH or NaH) which occurs at a bridging BHB to give a reactive anion containing a boron—boron bond, which is susceptible to electrophilic attack. Apart from the examples shown, based on the deprotonation of B_5H_9, tetraborane(10) can be used to form B_5H_{11}.

Carboranes are generally prepared[11,12] from boranes, often by reaction with alkynes, although reactions with acetylides can be used for the synthesis of monocarboranes. As with the boranes, thermal routes are important (and mechanistically complex).

The mechanism of isomerization of $1,2\text{-}C_2B_{10}H_{12}$ to the 1,7- and 1,12-isomers has been considered by several groups but remains the subject of debate. A number of routes are possible. For example, in the 'diamond–square–diamond' mechanism[2] two triangular faces of the cluster are combined to form a square which then reopens. Alternative ideas include the rotation of the three atoms within a triangular face or, more recently,[13] it has been proposed that clusters might rearrange via single edge cleavage. The only certainty about this problem would appear to be that it will not be solved overnight!

$$B_5H_9 + C_2H_2(\text{excess}) \xrightarrow{500\text{--}600°C} 2,4\text{-}C_2B_5H_7 + 1,6\text{-}C_2B_4H_6 + 1,5\text{-}C_2B_3H_5$$

$$B_5H_9 + RC\equiv CR \xrightarrow{Et_3N,25°C} 2,3\text{-}R_2C_2B_4H_6$$

$$B_8H_{12} + MeCCMe \longrightarrow (MeC)_2B_7H_9 + 5,6\text{-}(MeC)_2B_8H_{10}$$

$$B_{10}H_{14} + 2R_2S \longrightarrow B_{10}H_{12}(R_2S)_2 + H_2$$

$$B_{10}H_{14} + RC\equiv CR' \longrightarrow 1,2\text{-}(RC)(R'C)B_{10}H_{12}$$

$$\xrightarrow{450°C} 1,7\text{-isomer}$$

$$\xrightarrow{700°C} 1,12\text{-isomer}$$

$$LiC\equiv CMe + B_5H_9 \longrightarrow MeC\equiv CB_5H_9^- \longrightarrow MeCH_2CB_5H_7^- \xrightarrow{H^+}$$

$$MeCH_2CB_5H_8$$

The dicarbollide anions are obtained by degradation reactions similar to that used to prepare $B_9H_{14}^-$:

$$(HC)_2B_{10}H_{10} + EtO^- + 2EtOH \longrightarrow (HC)_2B_9H_9^- + B(OEt)_3 + H_2$$

Metal complexes can also be used in coupling or fusion reactions shown below.[14]

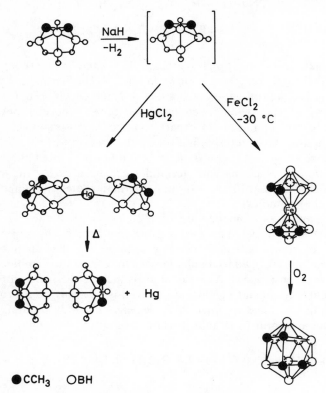

●CCH₃ ○BH

Finally, an interesting example of a cluster rearrangement from classical to 'Wadian' geometry was recently reported.[15] The $(RC)_4(BR')_6$ originally formed has an adamantane-like structure with bridging BR groups. However, on mild thermolysis it converts to a *nido* geometry.

closo   Adamantane–like   *nido*

●   ○
CCH₃ BC₂H₅

2.1.4 Physical Properties

The melting points of the *nido*-boranes, (Table 2.2) roughly parallel those of hydrocarbons of similar molecular weight, but because of the air sensitivity of the lower boranes vapour pressure measurements provide a more convenient method of assessing purity.

Their vibrational spectra have been thoroughly investigated and the most characteristic feature is the BH stretching vibrations. Terminal BH have $v(BH)$ 2460–2650 cm^{-1} and bridging BHB are seen at 1600–2100 cm^{-1}.

NMR spectroscopy is a very powerful technique when applied to boranes;[16] both ^1H and ^{11}B spectra are useful ($^{11}B, I = 3/2$, abundance 80%; $^{10}B, I = 3$, abundance 20%). Proton spectra are often broad as a result of the quadrapolar boron which increases the relaxation rate and consequently the line width of the signals. $^1H–^{11}B$ coupling is observed and ^1H signals split by a boron ($I = 3/2$) are seen as quartets (i.e. splitting is $2nI + 1$); Fig. 2.10 shows a typical example.

Table 2.2 Physical properties of some boranes

Compound	M.p./°C	B.p./°C
nido-Boranes:		
B_2H_6	− 165	− 93
B_5H_9	− 47	60
B_6H_{10}	− 62	108
B_8H_{12}	− 36	
$B_{10}H_{14}$	99.5	213 (extrap)
arachno-Boranes:		
B_4H_{10}	− 120	18
B_5H_{11}	− 122	65

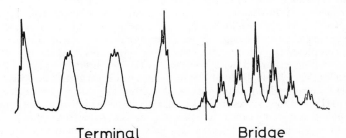

Terminal   Bridge

Figure 2.10 ^1H NMR spectrum of diborane(6). Couplings to ^{11}B ($I = 3/2$) result in four lines for the terminal BH_2 group and seven lines for the bridge hydrogen [as predicted using the simple ($nI + 1$) relationship]. The shoulders on some bands are due to couplings to ^{10}B. Reproduced with permission from S. Shore, in *Boron Hydride Chemistry* (Ed. E. L. Muetterties), Academic Press, New York, 1975.

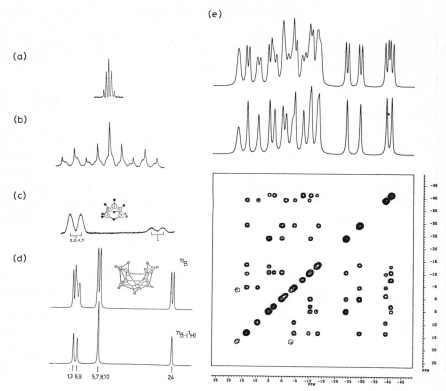

Figure 2.11 ^{11}B NMR spectra of (a) BH_4^-; (b) diborane(6); (c) *nido*-pentaborane(9); and (d) *nido*-decaborane(14), ^{11}B and ^{11}B–$\{^1H\}$. For, clarity, the spectra are *not* drawn with the same ppm scale. (a)–(c) are reproduced with permission from S. Shore in *Boron Hydride Chemistry* (Ed. E. L. Muetterties), Academic Press, New York, 1975; (d) courtesy of Professor N. N. Greenwood, University of Leeds. (e) 128 MHz ^{11}B NMR spectra of the *anti*-$B_{18}H_{21}^-$ anion (which is similar in structure to $B_{18}H_{22}$). The upper trace is the straightforward spectrum, the second an equivalent spectrum with complete $\{^1H\}$ decoupling. The *bottom* diagram is a 2D–$[^{11}B$–$^{11}B]$ COSY 90 contour plot (recorded with complete $\{^1H\}$ decoupling. The middle spectrum identifies the eighteen different ^{11}B resonance positions and, in comparison with the upper spectrum, identifies the sixteen BH (terminal) positions [doublets $^1J(^{11}B$–$^1H)$ ca 150 Hz) and the two resonances (singlets) due to the boron atoms common to the two *nido*-decaboranyl subclusters that notionally make up the eighteen-vertex anion. The off-diagonal COSY cross-correlations in the bottom diagram define nearest neighbour boron pairs, and sufficient correlations are present to assign the entire ^{11}B spectrum to the structure. Selective ^1H–$\{^{11}B\}$ can be used to relate the ^1H and ^{11}B spectra and a $[^1H$–$^1H]$ COSY experiment locates the precise positioning of the bridging hydrogen atoms around the faces of the anion. For a more detailed discussion, see X.L.R. Fontaine *et al.*, *J. Chem. Soc., Dalton Trans.*, 1988 (in press). Diagrams courtesy of X.L.R. Fontaine and J. D. Kennedy, Department of Inorganic and Structural Chemistry, University of Leeds.

As can be seen in the spectrum of diborane, terminal hydrogens ($\delta = 2$–7 ppm) have different chemical shifts to bridging hydrogens ($\delta = 0$–$^-$5 ppm) and larger couplings (terminal 150 Hz, bridge 50 Hz) to boron. The spin–spin couplings mean that spectra with overlapping signals are often obtained and so spectra are frequently measured with broad-band ^{11}B decoupling.

Boron NMR (Fig. 2.11) provides useful information about the number and type of groups present in a cluster. Usually ^{11}B spectra are measured since this nucleus is more abundant and of lower spin. The signals are split depending on the number of terminal borons present; thus a BH group is a doublet, a BH_2 group is a triplet, etc.. The interpretation of chemical shifts is non-trivial and, in the examples shown, unnecessary. Adequate information about the number and type of borons is often obtained from looking at the ^{1}H splittings. The spectra are simplified in that the splittings due to bridging BHB hydrogens and ^{11}B–^{11}B couplings are not normally observed because their magnitudes are less than the line widths of the spectra.

2.1.5 Reactions

These is an enormous wealth of chemistry associated with boranes[11,12] and we limit ourselves to selected reactions for diborane(6), pentaborane(9) and decarborane(14) shown in Figs 2.12–2.14.

Diborane(6) undergoes cleavage reactions with Lewis bases. Homolytic cleavage occurs with bulky amines, whereas with small amines asymmetric cleavage occurs. Substitution of terminal protons by alkyl groups or chlorine is readily accomplished. Reactions with alcohols and water cause complete degradation of the molecule wheareas with ammonia borazene (Chapter 4) is obtained. It is a good reducing agent, converting nitriles to amines; hydroboration proceeds anti-Markonikov and the resulting compounds can be converted into alkanes or alcohols and also undergo carbonylations, etc. We have already discussed reductive disproportionation as a route to higher boranes.

The reactivity of pentaborane(9) can be related to the calculated charge distribution in the molecule. Terminal hydrogens have more electron density than bridging hydrogens. The apical boron (1) has slightly higher electron density than the basal boron atoms—empirically this can be related to the number of hydrogen atoms that are withdrawing electrons from any particular boron. As a consequence, nucleophilic attack and deprotonation occur at basal borons and bridge hydrogens. The $B_5H_8{}^-$ anion may be isolated as a salt or used in a variety of reactions with substitution occurring at the base of the cluster to give, for example, conjuncto-2, 2'-$(B_5H_8)_2$. Friedel–Crafts and other electrophilic attacks occur at the apex, e.g. in the formation of 1-IB_5H_8 or in the synthesis of conjuncto-1, 2'-$(B_5H_8)_2$ from 2'-BrB_5H_8. The use of B_5H_9 and $B_5H_8{}^-$ in the preparation of metallaboranes is discussed on p. 31; their use in the synthesis of other boranes has already been illustrated (Fig. 2.9).

The reactivity of decaborane(14) can also be explained in the light of electron

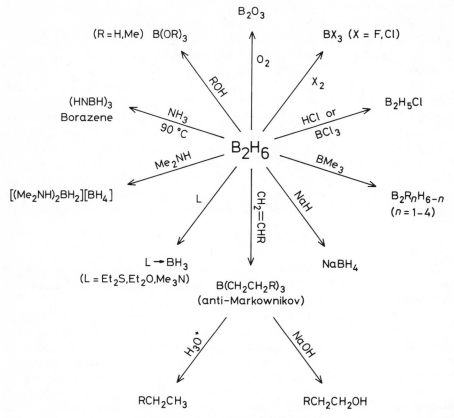

Figure 2.12 Reactions of diborane(6).

distribution within the molecule, the electron charge density decreases in the order $2, 4 > 1, 3 > 5, 7, 8, 10 > 6, 9$. We have already described some cluster expansion and degradation reactions used for the preparation of other boranes (p. 20). Treatment of $B_{10}H_{14}$ with bases gives the monoanion, formed by loss of a bridge proton, which can be isolated as, for example, the R_4N^+ salt. Use of excess sodium hydride gives the dianion $B_{10}H_{12}{}^{2-}$ whilst reaction with sodium reduces the cluster to *arachno*-$B_{10}H_{14}{}^{2-}$ which can be protonated to give $[B_{10}H_{15}]^-$. *Arachno* structures are also observed in the bis-adducts which are coordinated at positions 6 and 9. Formally, ligands such as CH_3CN, PR_3 and R_2S are two-electron donors and thus replacement of two protons by two of these ligands increases the cluster electron count by two causing a *nido* to *arachno* conversion. The Me_2S adducts are useful reagents and some examples are shown in Fig. 2.14. Alkyl derivatives can be prepared by Friedel–Crafts reactions.

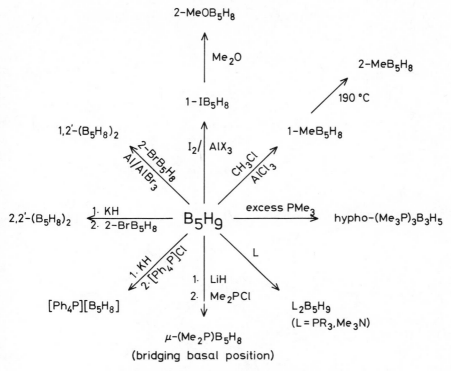

Figure 2.13 Reactions of *nido*-pentaborane(9).

Attempts to produce methyl derivatives give mixtures but reaction with ethyllithium followed by HCl gives a 90% yield of $6\text{-EtB}_{10}H_{13}$.

2.2 METALLABORANES

There is a wide range of compounds in which a BH or BH_2 group has been replaced by a transition metal fragment. The isolobal analogy allows us to predict which transition metal fragments can become 'honorary borons.' To be isolobal, two moieties must possess the same number of orbitals of similar geometry and energy and they should also contain the same number of electrons. The most common replacement is that of a BH group which provides three orbitals and two electrons for cluster bonding. With transition metals we have nine orbitals in total and it is assumed that the six lowest energy orbitals and their twelve electrons are non-bonding or involved in bonds to ligands. This leaves three orbitals together with their electrons for cluster bonding. To determine how many electrons a transition metal fragment can provide for cluster bonding, total all of the electrons around the metal and subtract 12—for convenience we shall generally

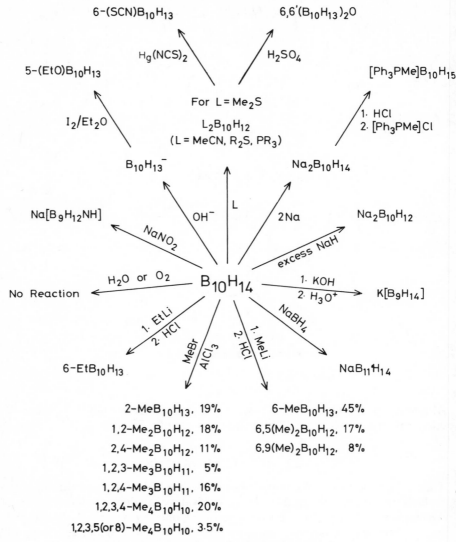

Figure 2.14 Reactions of *nido*-decaborane(10).

assume neutral fragments but, as will be indicated later, it is not especially important what oxidation state one assumes for the metal.

$$\text{Fe(CO)}_3 \ \ d^8 \ \ 8 + (3 \times 2) = 14 - 12 = 2 \ \ \text{BH}$$

The simplest metallaboranes contain the tetrahydroborate anion BH_4^- as a ligand; most examples have stabilizing ligands on the metal, although one of the earliest and best known examples, $U(BH_4)_4$, is an exception. This compound was prepared during efforts to obtain volatile uranium compounds for use in the separation of isotopes. Although the green solid is volatile above $30\,°C$ it was never used for this purpose; UF_6 is used nowadays.

$$UF_4 + 2Al(BH_4)_3 \longrightarrow U(BH_4)_4 + 2F_2AlBH_4$$

A wide range of transition metal tetrahydroborates are known.[17,18] Typically, they are prepared by salt elimination reactions of alkali metal tetrahydroborates with the appropriate halide complex:

$$2Cp_2TiCl_2 + 4LiBH_4 \longrightarrow 2Cp_2TiBH_4 + 4LiCl + B_2H_6 + H_2$$

$$(Ph_3P)_2CuCl_2 + 2NaBH_4 \longrightarrow (Ph_3P)_2CuBH_4 + 2NaCl$$

The structure of the above copper complex is very similar to diborane with effectively tetrahedral geometry at both copper and boron (Fig. 2.15). Addition of electrons to the system by increasing the number of ligands on the copper opens up the molecule to what is formally an *arachno* structure with only one bridging

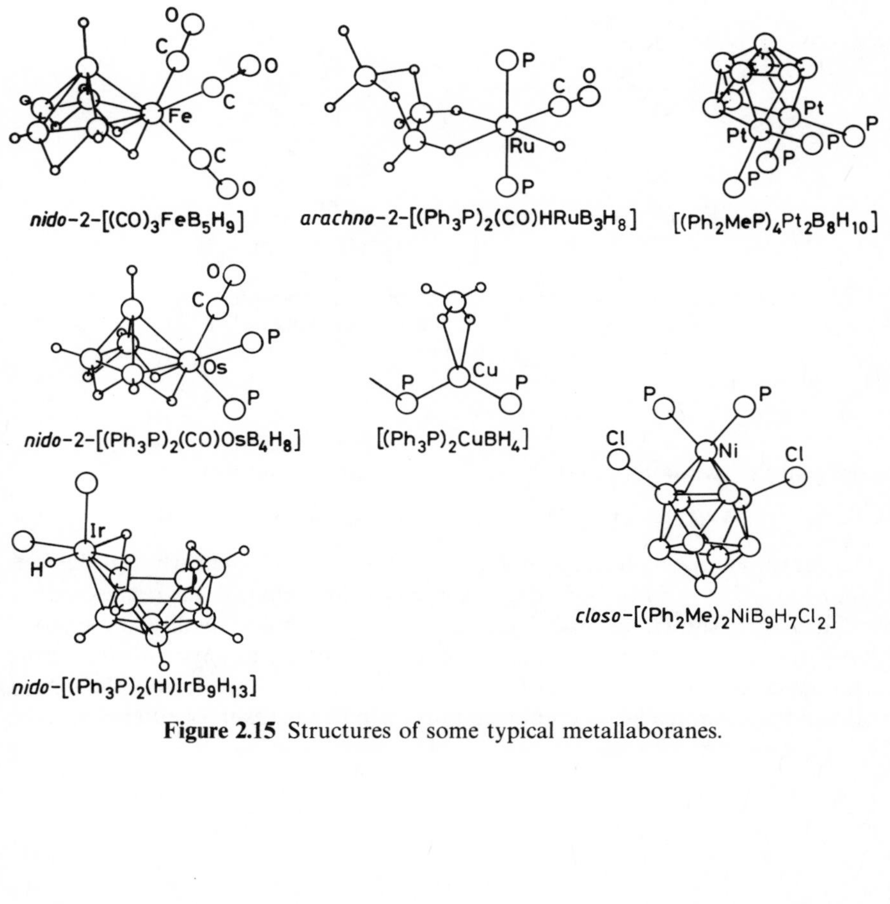

$nido\text{-}2\text{-}[(CO)_3FeB_5H_9]$　　$arachno\text{-}2\text{-}[(Ph_3P)_2(CO)HRuB_3H_8]$　　$[(Ph_2MeP)_4Pt_2B_8H_{10}]$

$nido\text{-}2\text{-}[(Ph_3P)_2(CO)OsB_4H_8]$　　$[(Ph_3P)_2CuBH_4]$　　$closo\text{-}[(Ph_2Me)_2NiB_9H_7Cl_2]$

$nido\text{-}[(Ph_3P)_2(H)IrB_9H_{13}]$

Figure 2.15 Structures of some typical metallaboranes.

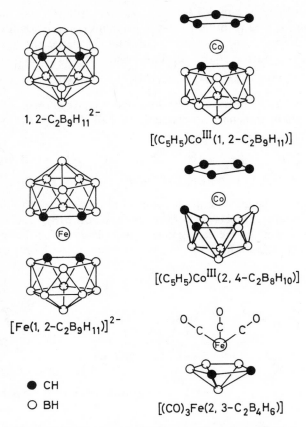

$1, 2\text{-}C_2B_9H_{11}{}^{2-}$

$[(C_5H_5)Co^{III}(1, 2\text{-}C_2B_9H_{11})]$

$[(C_5H_5)Co^{III}(2, 4\text{-}C_2B_8H_{10})]$

$[Fe(1, 2\text{-}C_2B_9H_{11})]^{2-}$

● CH
○ BH

$[(CO)_3Fe(2, 3\text{-}C_2B_4H_6)]$

Figure 2.16 Structures of some metallacarboranes. The structure of $1, 2\text{-}C_2B_9H_{11}{}^{2-}$ is drawn showing its five sp^3 orbitals.

hydrogen; the two compounds are in equilibrium in solution.

$$nido\text{-}(Ph_3P)_2CuBH_4 + PPh_3 = arachno\text{-}(Ph_3P)_3CuBH_4$$

An enormous array of compounds containing borane[19] and carborane ligands are now known[20] and below we give some typical preparations; representative structures are shown in Figs 2.15 and 2.16. The most readily understood preparative routes are metathetical reactions of borane anions with transition metal halide complexes (often accompanied by oxidative insertion of the metal). Many of the other reactions employed appear to be the result of either genius or inspired guesswork!

$$NbCl_5 + Li(C_5Me_5) + NaBH_4 \longrightarrow (C_5Me_5)_2Nb_2(B_2H_6)$$

$$Mn(CO)_5Cl + NaB_3H_8 \longrightarrow arachno\text{-}(OC)_4MnB_3H_8 + NaCl + CO$$

$$\xrightarrow{\text{heat or } hv} nido\text{-}(CO)_3MnB_3H_8 + CO$$

$$Ru(CO)ClH(PPh_3)_3 + TlB_3H_8 \longrightarrow$$

$$arachno\text{-}Ru(B_3H_8)(CO)H(PPh_3)_2 + TlCl + PPh_3$$

$$nido\text{-}Os(B_5H_9)(CO)(PPh_3)_2 \xrightarrow{\text{heat}}$$

$$nido\text{-}Os(B_4H_8)(CO)(PPh_3)_2$$

$$Fe(CO)_5 + nido\text{-}B_5H_9 \longrightarrow Fe(B_4H_8)(CO)_3$$

$$Fe(CO)_5 \text{ (excess)} + nido\text{-}B_5H_9 + LiAlH_4 \longrightarrow$$

$$nido\text{-}[Fe(CO)_3](B_3H_7)1\%$$

$$Ir(CO)Cl(PPh_3)_2 + KB_5H_8 \longrightarrow$$

$$nido\text{-}(Ph_3P)_2(CO)Ir(B_5H_8) + KCl$$

$$Ru(CO)ClH(PPh_3)_3 + [Et_4N][B_9H_{14}] \longrightarrow$$

$$nido\text{-}Ru(CO)(PPh_3)_2(B_9H_{13})$$

$$PtCl_2(PMe_2Ph)_2 + [Et_4N][B_9H_{14}] \longrightarrow$$

$$nido\text{-}Pt(B_8H_{12})(PMe_2Ph)_2 \xrightarrow[\text{2.PtCl}_2(PMe_2PH)_2]{\text{1.2KH}}$$

$$arachno\text{-}\{Pt(PMe_2Ph)_2\}_2(B_8H_{12})$$

$$C_2B_9H_{11}{}^{2-} + C_5H_5{}^- + Fe^{2+} \longrightarrow FeCp(C_2B_9H_{11})$$

Chemically, metallaboranes are often substantially more thermodynamically and air stable than their isostructural parent boranes and some rhodium carboranes have been used as catalysts.

We have already indicated how simple electron counting rules may be applied to transition metal fragments and thus enable the prediction of geometries. One difficulty which is often encountered is in deciding the oxidation state of the metal. In practice it must be remembered that formal oxidation numbers are frequently related to convention more than to chemical properties. For example, if we consider the electron count in $Os(B_4H_8)(CO)(PPh_3)_2$, we could consider the osmium to have oxidation state II or 0. In the former case, on counting electrons we have a metal fragment $Os^{II}(CO)(PPh_3)_2$, which provides $6 + 2 + (2 \times 2) - 12$

$= 0$ and a borane fragment $B_4H_8{}^{2-}$, which provides $(4 \times 3) + (8 \times 1) + 2 - (4 \times 2) = 14e^-$ for cluster bonding. The total of 7 electron pairs is $n + 2$ pairs for the five-vertex cluster and hence the predicted strucure is *nido*. It is clear that if we assume that osmium has oxidation state 0 (and provides $2e^-$) then the borane fragment will be neutral and provide only $12e^-$, leading to an identical total electron count and the same predicted structure as before. Whilst not all situations are as simple as the above, it does highlight the relative unimportance of formal oxidation state/number. Wherever possible it is useful to assign oxidation number on the basis of a physical property such as NMR coupling constants.

Some difficulties can arise from the over-optimistic assumption that metal vertices have to provide three orbitals for cluster bonding, and ref. 21 illustrates one current controversy. For a discussion of the extension of polyhedral skeletal electron pair theory to non-conical fragments, the recent work of Mingos and co-workers[22,23] is invaluable. The isolobal principle is discussed in some detail in Hoffmann's Nobel Lecture.[24]

The chemistry of metallacarboranes predates that of boranes by a number of years and, although the structures of many of the clusters can be rationalized using the same principles, historically the preparations were founded on different preconceptions. The description of the bonding in $B_9C_2H_{11}{}^{2-}$ has many similar features to that used for the cyclopentadienyl ligand in ferrocene. Hawthorne and co-workers[25] deduced that a class of compounds based on this analogy could be prepared and went on to synthesize a wide range of materials including rhodacarboranes, which appear to have useful catalytic properties in homogeneous hydrogenation and the isomerization of alkenes. Some examples are shown in Fig. 2.16. More recently, work on 'carbon-rich' carboranes has also been discussed.[26]

2.3 GROUP IV AND V IONS

2.3.1 Zintyl Anions

It is almost 100 years since Joannis reported that solutions of sodium in liquid ammonia apparently reduced lead to give intense green materials. A number of workers have investigated these types of reactions, in particular Zintyl, who during the 1930s carried out detailed potentiometric titrations and established the existence, in solution, of tin, lead, antimony and bismuth anions such as $Sn_9{}^{4-}$, $Pb_7{}^{4-}$ and $Bi_5{}^{3-}$.

Attempts to obtain solid compounds (e.g. by evaporation of liquid ammonia) result in alloys, i.e. the anions formed:

$$Sb_7{}^{3-}/liq.\ NH_3 \longrightarrow [Na(NH_3)_n]_3Sb_7$$

$$\downarrow$$

$$4Sb(s) + 3NaSb_3(s) + 3nNH_3(g)$$

are themselves very good reducing agents and attack the alkali metal cation, partly because the alloys formed are relatively stable materials. Corbett[27] recognized the difficulty and solved this problem by sequestering the cation with a bulky ligand, 2, 2, 2-crypt. This approach has been extremely successful and has led to the isolation of many anions. A typical reaction is shown below and the range of known compounds is shown in Table 2.1 with selected structures in Fig. 2.17. Heteropoly anions are prepared in a similar manner.

$$NaSn_{2.25}(alloy) + crypt + ethylenediamine \longrightarrow [cryptNa]_4Sn_9$$

There are a number of isoelectronic relationships between Zintyl anions, boranes and other main group systems, for example $Sb_4^{2-} \equiv Te_4^{2+}$, $Sn_9^{2-} \equiv Ge_9^{2-} \equiv B_9H_9^{2-} \equiv Bi_9^{5+}$ and it is interesting to compare structures within these groups and to consider if Wade's rules can be successfully applied.

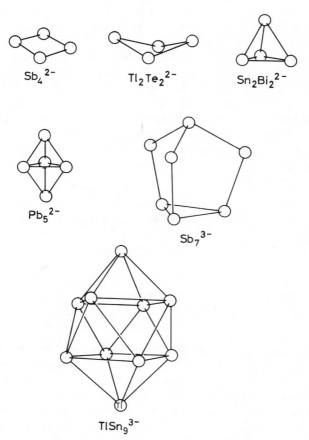

Figure 2.17 Structures of some Zintyl anions.

In Sb_4^{2-} the total electron count is $22\,e^-$ and the number of electron pairs available for cluster bonding is therefore 11, i.e. $n + 3$. This anion should therefore have an *arachno* structure, derived from an octahedron, similar to B_4H_{10}. The observed structure is actually square planar. Wade's rules have apparently failed, *but* as the structure obtained is derived from an octahedron by loss of different vertices to those used to explain the structure of B_4H_{10}, we might thus identify the anion as *iso-arachno*-Sb_4^{2-}. This species is isostructural with S_4^{2+} (p. 114).

In the $20\,e^-$ clusters $Tl_2Te_2^{2-}$ and $Sn_2Bi_2^{2-}$, with $n + 2\,e^-$ pairs for cluster bonding, the predicted structure is *nido* by loss of a vertex from a trigonal bipyramid. The butterfly structure of $Tl_2Te_2^{2-}$ is hence predicted whereas the tetrahedral geometry in $Sn_2Bi_2^{2-}$ makes this an *iso-nido* cluster!

The structure of Sb_7^{3-} presents severe problems. This anion is isoelectronic and isostructural with P_4S_3 and P_7^{3-} and is readily explicable using Gillespie's system as a triply edge bridged tetrahedron. If we are determined, it is still possible to force the anion partially to fit Wade's rules by ignoring the electrons on the bridging Sb^- atoms, leaving $20\,e^-$ on the tetrahedral skeleton—this then has an *iso-nido* geometry.

This last argument only serves to emphasize the difficulties of over-enthusiastic use of empirical methods. These treatments provide a 'framework of expectation' and are extremely useful in helping to systematize and simplify an apparent jungle of structures. However, it is unwise to expect them to be applicable in every case unless we accept a heavy burden of additional guidelines, which would only serve to make the methods too cumbersome for general use.

As a final illustration of the difficulties in this area, it is worth considering some nine-vertex clusters:

$B_9H_{12}^-$	$22\,e^-$	C_{4v}	— mono-capped square antiprism (predicted *nido* structure)
Sn_9^{4-}	$22\,e^-$	C_{4v}	— mono-capped square antiprism (predicted *nido* structure)
Bi_9^{5+}	$22\,e^-$	D_{3h}	— tricapped trigonal prism 'wrong' structure
$B_9H_9^{2-}$	$20\,e^-$	D_{3h}	— tricapped trigonal prism (predicted *closo* structure)
Sn_9^{3-}	$21\,e^-$	D_{3h}	— odd-electron species (no structural prediction!)

As can be seen above, although the tin anions generally obey Wade's rules, the isoelectronic species Bi_9^{5+} is rule-breaking. A closer look at these two structures reveals that only a small distortion is needed to transform between them, although close examination of models is needed before the structural similarities are appreciated. Figure 2.18 shows the two polyhedra and the atom movements required in their transformation; this example has been studied in some detail.[27] The ease of change between different cluster geometries is evident in the

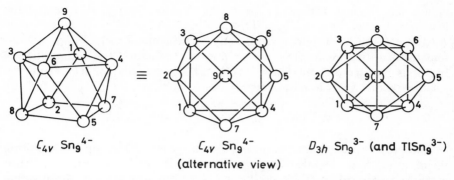

C_{4v} Sn$_9^{4-}$ C_{4v} Sn$_9^{4-}$ D_{3h} Sn$_9^{3-}$ (and TISn$_9^{3-}$)

(alternative view)

Figure 2.18 Structures Sn$_9^{4-}$ and Sn$_9^{3-}$ showing the similarity between C_{4v} and D_{3h} symmetry.

fluxionality of many species and further emphasizes the limitations of empirical structural relationships.

NMR studies are a useful probe of the structure of species in solution as well as giving information about the fluxional processes. For example, the Sn$_9^{4-}$ anion has only a single resonance (split by coupling to ^{117}Sn) in its ^{119}Sn spectrum— clear evidence of rapid rearrangement of the clusters.[28] The detection of compounds that are unstable or difficult to isolate is also possible. The reaction[29] of Pt(PPh$_3$)$_4$ with K$_4$Sn$_9$ in ethylenediamine has been followed by a combination of ^{31}P and ^{119}Sn NMR and *nido*-(PPh$_3$)$_2$Pt(Sn$_9$), a rare example of a cluster containing transition and main group metals, is postulated as the product. Mixed tin–lead–thallium anions have also been detected.[30]

Finally, before leaving this topic it is worth mentioning that the solid-state structures of alkali metal silicides often contain anions with now familiar geometries. The relationship between solid-state and 'molecular' structures has been neatly demonstrated in a review.[31,32] Interestingly, silicon anions include Si$_4^{6-}$ and Si$_4^{4-}$, which have the predicted *arachno* butterfly and *nido* tetrahedral structures, respectively.

2.3.2 Bismuth Cations

Structurally related to the above compounds are cationic bismuth clusters, Bi$_8^{2+}$, Bi$_9^{5+}$ and Bi$_5^{3+}$. In contrast to the Zintyl anions, they are most often obtained[33,34] from melts and the tendency for disproportionation is often greatly reduced by the use of a large non-oxidizing anion of low basicity (e.g. AlCl$_4^-$). Bi$_8^{2+}$ has a square antiprismatic structure which can be regarded as an *iso-arachno* structure based on B$_{10}$H$_{10}^{2-}$. Bi$_9^{5+}$ (mentioned above) is actually present in a compound of formula Bi$_6$Cl$_7$ together with BiCl$_5^{2-}$ and Bi$_2$Cl$_8^{2-}$ anions. The X-ray structure of Bi$_5$(AlCl$_4$)$_3$ has recently been reported[35] and the cation has a *closo* trigonal bipyramidal structure similar to that of the isoelectronic Pb_5^{2-}.

Although the preparations are usually carried out in melts, simple stoichiometric quantities of materials are not always satisfactory. The optimization of reactions often requires great care and effort. The preparation of Bi_5^{3+} requires a slight excess of $AlCl_3$ to ensure acidity of the melt:

$$3.5Bi + BiCl_3 + 3.1AlCl_3 \xrightarrow[350°C]{NaAlCl_4} Bi_5(AlCl_4)_3$$

Some interesting examples at the interface between main group and transition metal clusters have recently been prepared for bismuth[36,37] and antimony.[38] For example, treatment of $[Et_4N][BiFe_3(CO)_{10}]$ with CO at moderate pressure gives $[Et_4N]_2[Bi_4Fe_4(CO)_{13}]$ together with $Fe(CO)_5$. The structure of the dianion consists of a Bi_4 tetrahedron with an apical $Fe(CO)_4$ unit and three faces bridged by $Fe(CO)_3$ groups. Other bismuth–transition metal clusters include $[Bi_2Fe_2Co(CO)_{10}]^-$ and $[Bi_2Fe_4(CO)_{13}]$. It seems likely that a variety of structures will come to light in these types of system.

Antimony may also be stabilized as part of a metal carbonyl cluster, as shown in Fig. 2.19.

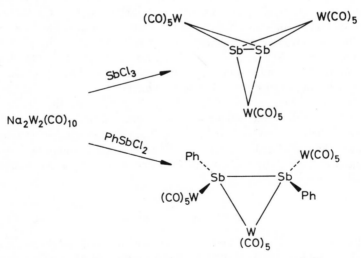

Figure 2.19 Stabilization of antimony in transition metal clusters.

2.4 TRANSITION METAL CLUSTERS

We have already indicated how metals can be involved as part of a borane cluster. There is also an enormous number of transition metal clusters.[4,39,40] Many of the

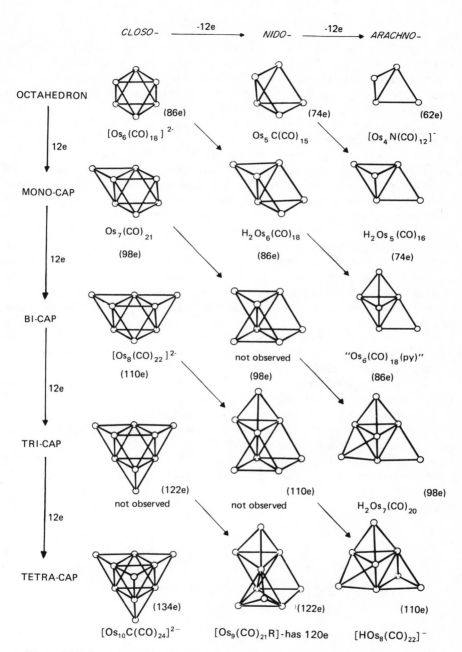

Figure 2.20 Some metal carbonyl clusters. Apart from the familar *closo–nido–arachno* relationships (on going from left to right), the structures lying along the diagonals represent alternative geometries for the same electron count resulting from the removal of a vertex and its replacement by a capping group. Reproduced with permission from M. McPartlin, *Polyhedron*, 1984, 3, 1279.

known compounds are organometallic with carbonyl, cyclopentadienyl or related ligands. Transition metal clusters may provide useful models for heterogeneous catalysts and so, apart from their intrinsic academic interest, they also offer scope for the study of important processes such as Fischer–Tropsch synthesis.

In general, smaller clusters tend to have structures which can be related to Wade's rules whilst larger (high nuclearity) clusters are more readily rationalized on the basis of metallic close packing. We have already mentioned how the number of electrons available for cluster bonding is calculated; in summary, the number of valence electrons is totalled and $12\,e^-$ per metal are subtracted. Thus $Co_4(CO)_{12}$ has a total of 60 valence e^- and is considered to have $12\,e^-$ available for cluster bonding; the predicted *nido* structure is observed.

Although it is possible to predict some of the known structures using Wade's rules alone, the diversity of structures is such that additional ideas are needed. Some very sophisticated treatments have been attempted,[41,42] but there seems little advantage in them since they are very complex to master without being particularly helpful in explaining bonding. At present, the relatively straightforward approaches of Mingos[43,44] and McPartlin[45] seem to be adequate for large and small clusters. Figure 2.20 illustrates the relationship between a variety of compounds incorporating the familiar *closo*, *nido* and *arachno* concepts and the structures obtained via edge sharing and face bridging. Figure 2.21 shows examples of structures based on Pt_3 triangles which are stacked helically, whilst Fig. 2.22 contains examples of clusters whose structures can be related to metal 'close-packed' systems based on, for example, hexagonal and cubic close packing. Examples of mixed metal clusters can be found in ref. 46.

The above examples are essentially organometallic clusters. One important class of clusters are the cubanes, of which the best known are probably ferrodoxin and nitrogenese. A considerable volume of work on understanding the structures and properties of the metal–sulphur cluster cores in these biologically

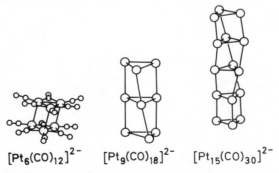

$$[Pt_6(CO)_{12}]^{2-} \qquad [Pt_9(CO)_{18}]^{2-} \qquad [Pt_{15}(CO)_{30}]^{2-}$$

Figure 2.21 Structures of some helical platinum carbonyl clusters. Reproduced by permission of the publisher.

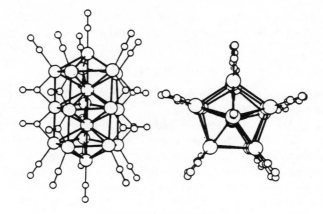

$[Pt_{19}(\mu\text{-}CO)_{10}(CO)_{12}]^{4-}$

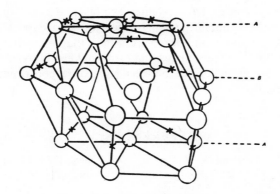

$[Pt_{26}(\mu\text{-}CO)_{9}(CO)_{23}]^{2-}$

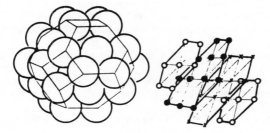

$[Pt_{38}(CO)_{44?}]^{2-}$

Figure 2.22 Structures of some high-nuclearity platinum carbonyl clusters. Reproduced with permission from D. M. P. Mingos and R. W. M. Wardle, *Transition Met. Chem.*, 1985, **10**, 441.

important molecules has been carried out over the past few years.[47] One strategy that has been adopted is the synthesis of model systems.[48]

2.5 BORON CHLORIDES

An interesting group of compounds which do not obey Wade's rules particularly well are the polyhedral boron chlorides. They have the general formula B_nCl_n in which those with $n = 4, 8$ and 9 have been characterized crystallographically and those with $n = 11-20$ are also known spectroscopically. B_4Cl_4 has a tetrahedral structure whereas B_8Cl_8 and B_9Cl_9 are isostructural with $B_8H_8{}^{2-}$ and $B_9H_9{}^{2-}$. All of the chloro compounds contain only $2n$ cluster bonding electrons and their properties pose a number of problems. For example, why do they exist when the analogous neutral boranes are not known? In general they are relatively thermally stable; B_9Cl_9 decomposes only slowly as a solution in BCl_3.

A recent theoretical and photoelectron spectroscopic study[49] has attempted to address some of the difficulties. The tangential orbitals of the chlorine atoms are important. Back-bonding from filled Cl 3p orbitals into unoccupied B 2p atomic orbitals occurs, but in a more complex fashion than that encountered in BCl_3. The LUMO of the B_nCl_n clusters corresponds approximately to the HOMO of the $B_nH_n{}^{2-}$ species, but in the former there is a substantial contribution derived from Cl atomic orbitals.

2.6 REFERENCES

1. J. B. Casey, W. J. Evans and W. H. Powell, *Inorg. Chem.*, 1983, **22**, 2228 and 2236.
2. W. N. Lipscomb, *Boron Hydrides*, Benjamin, New York, 1963.
3. K. Wade, *Adv. Inorg. Radiochem.*, 1976, **18**, 1.
4. B. F. G. Johnson (Ed.), *Transition Metal Clusters*, Wiley, Chichester, 1980.
5. A. J. Stone, *Polyhedron*, 1984, **3**, 1299.
6. A. Stock, *Hydrides of Boron and Silicon*, Cornell University Press, New York, 1933.
7. H. C. Brown, *Organic Syntheses via Boranes*, Wiley, New York, 1975.
8. R. T. Holzmann (Ed.), *Production of Boranes and Related Research*, Academic Press, New York, 1967.
9. R. Greatrex, N. N. Greenwood and C. D. Potter, *J. Chem. Soc., Dalton Trans.*, 1986, 81.
10. S. G. Shore, in *Rings, Clusters and Polymers of the Main Group Elements* (Ed. A. H. Cowley), ACS Symposium Series, No. 232, Chapter 1, pp. 1–16 American Chemical Society, Washington, DC 1982.
11. E. L. Muetterties (Ed.), *The Chemistry of Boron and Its Compounds*, Wiley, New York 1967.
12. E. L. Muetterties (Ed.), *Boron Hydride Chemistry*, Academic Press, New York, 1975.
13. B. F. G. Johnson, *J. Chem. Soc., Chem. Commun.*, 1986, 27.
14. R. N. Grimes, *Acc. Chem. Res.*, 1983, **16**, 22.
15. R. Koster, G. Siedel and B. Wrackmeyer, *Angew. Chem., Int. Ed. Engl.*, 1985, **24**, 326.
16. G. R. Eaton and W. N. Lipscomb, *NMR Studies of Boron Hydrides and Related Compounds*, Benjamin, New York, 1969.

17. B. D. James and M. G. H. Wallbridge, *Prog. Inorg Chem.*, 1970, **11**, 99.
18. T. J. Marks and J. R. Kolb, *Chem. Rev.* 1977, **77**, 263.
19. J. D. Kennedy, *Prog. Inorg. Chem.*, 1984, **32**, 519.
20. R. N. Grimes (Ed.), *Metal Interactions with Boron Clusters*, Plenum Press, New York, 1982.
21. R. T. Baker, *Inorg. Chem.*, 1986, **25**, 109; J. D. Kennedy, *Inorg. Chem.*, 1986, **25**, 111; R. L. Johnston and D. M. P. Mingos, *Inorg. Chem.*, 1986, **25**, 3321.
22. D. G. Evans and D. M. P. Mingos, *Organometallics*, 1983, **2**, 435.
23. D. M. P. Mingos *Acc. Chem. Res*, 1984, **17**, 311.
24. R. Hoffmann, *Science*, 1981, **211**, 995; *Angew. Chem., Int. Ed. Engl.*, 1982, **21**, 711.
25. M. F. Hawthorne and co-workers, *J. Am. Chem. Soc.*, 1984, **106**, 2965, 2979, 2990, 3004 and 3011.
26. R. N. Grimes, *Adv. Inorg. Radiochem.*, 1983, **26**, 55.
27. J. D. Corbett, *Chem. Rev.*, 1985, **85**, 383.
28. R. W. Rudolph, W. L. Wilson, F. Parker, R. C. Taylor and D. C. Young, *J. Am. Chem. Soc.*, 1978, **100**, 4629.
29. F. Teixidor, M. L. Luetkens, Jr, and R. W. Rudolph, *J. Am. Chem. Soc.*, 1983, **105**, 149.
30. W. L. Wilson, R. W. Rudolph, L. L. Lohr, R. C. Taylor and P. Pyykko, *Inorg. Chem.*, 1986, **25**, 1535.
31. H. G. von Schnering, *Angew. Chem., Int. Ed. Engl.*, 1981, **20**, 33.
32. H. G. von Schnering, M. Schwarz and R. Nesper, *Angew. Chem., Int. Ed. Engl.*, 1986, **25**, 566.
33. R. J. Gillespie and J. Passmore, *Adv. Inorg. Radiochem.*, 1975, **17**, 49.
34. J. D. Corbett, *Prog. Inorg. Chem.*, 1976, **21**, 129.
35. B. Krebs, M. Mummert and C. Brendel, *J. Less-Common Met.*, 1986, **116**, 159.
36. K. H. Whitmire, M. R. Churchill and J. C. Fettinger, *J. Am. Chem. Soc.*, 1985, **107**, 1056.
37. K. H. Whitmire, K. S. Raghuveer, M. R. Churchill, J. C. Fettinger and R. F. See, *J. Am. Chem. Soc.*, 1986, **108**, 2778.
38. G. Huttner, U. Weber, B. Sigwarth and O. Scheidsteger, *Angew. Chem., Int. Ed. Engl.*, 1982, **21**, 215.
39. G. Schmid, *Struct. Bonding*, 1985, **62**, 51.
40. M. McPartlin and D. M. P. Mingos, *Polyhedron*, 1984, **3**, 1321.
41. B. K. Teo and N. J. A. Sloane, *Inorg. Chem.*, 1985, **24**, 4545.
42. B. K. Teo, *Inorg. Chem.*, 1984, **23**, 1251.
43. D. M. P. Mingos, *J. Chem. Soc., Chem. Commun.*, 1985, 1352.
44. D. M. P. Mingos and R. W. M. Wardle, *Transition Met. Chem.*, 1985, **10**, 441.
45. M. McPartlin, *Polyhedron*, 1984, **3**, 1279.
46. M. J. McGlinchey, M. Meekuz, P. Bougeard, B. G. Sayer, A. Marinetti, J. Y. Saillard and G. Jauven, *Can. J. Chem.*, 1983, **61**, 1319.
47. A. J. Thomson, in *Metalloproteins, Part 1: Metal Proteins with Redox Roles* (Ed. P. M. Harrison), Macmillan, London, 1985, pp. 79–120.
48. B. A. Averill, *Struct. Bonding*, 1983, **53**, 59.
49. P. R. LeBreton, S. Urano, M. Shahbaz, S. L. Emery and J. A. Morrison, *J. Am. Chem. Soc.*, 1986, **108**, 3937.

CHAPTER THREE

Electron-Precise/Classical Species

3.1 NEUTRAL SULPHUR AND MIXED SULPHUR–SELENIUM RINGS

Most chemists know of the many allotropes that exist for elemental sulphur[1] and are familiar with the crown structure of *cyclo*-octasulphur, S_8, but are unaware that this is only one member of a homologous series of S_n ($n = 6$–26) rings.[2] Additionally, cyclic sulphur oxides[3] such as S_6O and S_7O_2 and mixed sulphur–selenium systems[4] are known (Table 3.1). The latter compounds are of some commercial significance, with applications as diverse as anti-dandruff shampoos, fireworks and polymerization inhibitors.

The preparation of S_n rings can be accomplished by a variety of routes:

$$Na_2S_2O_3 + 2HCl(aq.) \longrightarrow 1/nS_n + SO_2 + 2NaCl + H_2O$$

$$S_2Cl_2 + 2KI \longrightarrow S_2I_2 + 2KCl \longrightarrow 1/nS_{2n} + I_2$$

Acid hydrolysis of thiosulphate was originally used for the preparation of S_6, which is separated from S_7 and S_8 by recrystallization from toluene. A less hazardous method is the thermal decomposition (at room temperature) of diiodosulphane, S_2I_2, (formed *in situ*) which gives a 36% yield of S_6 and minor amounts of higher *cyclo*-sulphanes after purification. S_6 is obtained as orange–yellow crystals which readily decompose at room temperature. Some of the larger

Table 3.1 The known neutral Group VI rings; the indicated symmetries are for the parent S_n rings

Ring size	Species		Symmetry
S_6	S_6O	Se_6, Se_5S, Se_4S_2	D_{3d}
S_7	S_7O, S_7O_2	Se_7, Se_5S_2	C_1
S_8	$S_8O, (S_8O)_2SnC_4$	$Se_{8-n}S_n$ ($n = 0$–7)	C_2
S_9	S_9O		C_1
S_{10}	$S_{10}O$		C_2
S_{11}			C_1
S_{12}	$S_{12}O_2$	$Se_{12-n}S_n$	C_{2h}
S_{13}			C_1
S_{18}			C_1
S_{20}			C_2

rings are more stable and so can be prepared from molten S_8. For example, if commercial sulphur is heated to 200 °C, cooled to *ca* 150 °C and finally quenched by pouring into liquid nitrogen a mixture of S_7, S_{12}, S_{18} and S_{20} rings is obtained. The individual compounds are only obtained after painstaking recrystallizations, however. An alternative route to S_{18} is the reaction of $S_{10}Cl_2$ with H_2S_8, but in this case the preparation of the starting materials is very tedious (and occasionally unpleasant!).

Fortunately, rational syntheses based on reactions of Cp_2TiS_5 with chlorosulphanes (and consequent elimination of Cp_2TiCl_2) have been developed (Fig. 3.1). Those readers interested in the application of the isolobal principle (cf. p. 28) will of course recognize that the Cp_2Ti fragment is isolobal with the sulphur atom and therefore this compound is actually a stable analogue of S_6. Recently, this approach has been extended to permit the formation of Se_5S, Se_5S_2 and Se_7 by treatment of Cp_2TiSe_5 with SCl_2, S_2Cl_2 and Se_2Cl_2, respectively.[5] The use of transition metal complexes as synthetic reagents in this way will no doubt become more widespread in the future. Surprisingly, the preparation of S_6 from Cp_2TiS_5 is not the best route (probably because SCl_2 is difficult to obtain pure).

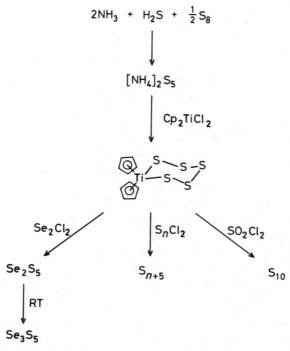

Figure 3.1 Preparation of S_n and S_nSe_m rings from Cp_2TiS_5.

The preferred route is reaction of S_2Cl_2 with H_2S_4:

$$S_2Cl_2 + H_2S_4 \longrightarrow cyclo\text{-}S_6 + 2HCl$$

85% yield

Eight membered and larger sulphur–selenium rings may be obtained from melts of the elements, by reaction of chlorosulphanes with hydrogen selenide (or *vice versa*), from selenous acid or by decomposition of mixtures of Se_2I_2 and S_2I_2, although many of these reactions give rise to mixtures of isomers:

$$Se_2Cl_2 + H_2S_n \longrightarrow Se_nS_{12-n} + 2HCl$$

$$2H_2S + SeO_2(aq.) \longrightarrow Se_nS_{8-n}(\downarrow) + 2H_2O$$

$$S_2Cl_2/Se_2Cl_2 + KI \longrightarrow S_2I_2/Se_2I_2 + KCl \longrightarrow Se_nS_{8-n} + I_2$$

The cyclic sulphur oxides are normally prepared by oxidation of their parent sulphur rings using *per*fluoroacetic acid, although when S_8 is treated with excess acid S_7O_2 and SO_2 are obtained, suggesting that ring contraction proceeds *via* a currently unknown oxidized eight-membered ring:

$$S_n + CF_3CO_3H \longrightarrow S_nO + CF_3CO_2H \ (n = 6\text{–}10)$$

$$S_n + 2CF_3CO_3H \longrightarrow S_nO_2 + CF_3CO_2H \ (n = 7, 9)$$

Using our original definition (p. 2) for the number of electrons involved in 'cluster' bonding S_n rings are electron-rich with a total of $6n$ electrons (with $4n$ available for 'cluster bonding') and should therefore have open, not cage, structures. As can be seen in Fig. 3.2, all of the species consist of puckered rings with variable bond angles and distances. High-symmetry rings such as S_8 and S_6 have normal S—S bond distances (*ca* 2.05 Å) and torsional angles (*ca* 85°); however lower symmetry species can be distorted. The most interesting example in this regard is S_7, which would be expected to have C_2 symmetry like cycloheptane but actually has C_s symmetry with four of the sulphur atoms in a plane and consequently one torsional angle of close to 0°. The S—S distances for the four coplanar sulphur atoms are unusual and may imply some π bonding contributions (Fig. 3.3).

An empirical rationalization of the structure of S_7 has been provided by Gillespie[6] (Fig. 3.4). A convenient starting polyhedron is a cube. This has one edge bridged, two vertices removed and finally the addition of a pair of electrons to break one of the basal bonds. Figure 3.4 also shows how the structures of S_7O and S_7I^+ may be related to that of S_7. It is worth noting that our choice of a cube as a starting point is not exclusive—the structure of S_7 can be derived from a six-vertex trigonal prism and this is a useful exercise for the interested reader. Similarly, it is possible to derive the observed chair form, cyclohexane-like, structure of S_6 (Fig. 3.4b).

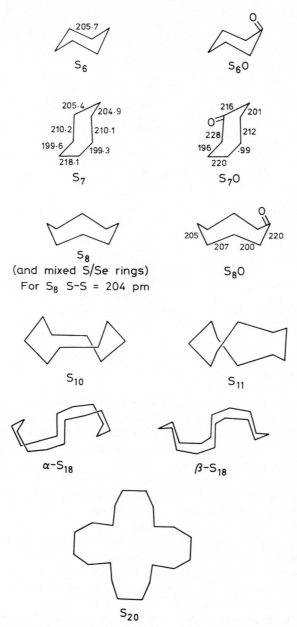

Figure 3.2 Structures of the known neutral homocyclic S_n rings. Mixed sulphur–selenium rings are also known. Bond lengths in pm.

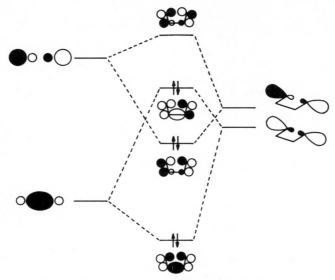

Figure 3.3 MO scheme showing the interaction between the S—S σ orbital and the lone pairs on adjacent sulphur atoms in S_7.

As mentioned above, the S_n rings are formally 'electron-rich,' but this only serves to highlight the difficulties of these empirical treatments and other arbitrary labels. Cyclohexane and S_6 are isoelectronic (in valence electron terms) and it is unlikely that any organic chemist would consider either compound to have any excess of electrons! The solution to the problem lies in considering the number of electrons (either as lone pairs or in bonds to exocyclic atoms) *not* involved in cluster bonding. Careful examination of Fig. 3.4b reveals that each atom in S_6 has two non-bonding lone pairs—the equivalent in cyclohexane being the two C—H bonds on each carbon. In Fig. 3.4 we started with an electron-precise ($5n$) cluster with one lone pair on each vertex; after manipulation we have obtained an 'electron-rich' six-membered ring with two lone pairs on each vertex. An organic chemist would have started by assuming that each atom is sp³ hybridized and in consequence would have been able to suggest the cyclohexane structure. Hence the structural conclusions reached by organic chemists and those obtained by cluster chemists are identical (in this case), even though totally different starting points are used.

The structures of mixed sulphur–selenium rings are analogous to their sulphur analogues. Structures in which the number of S—Se bonds is minimized appear

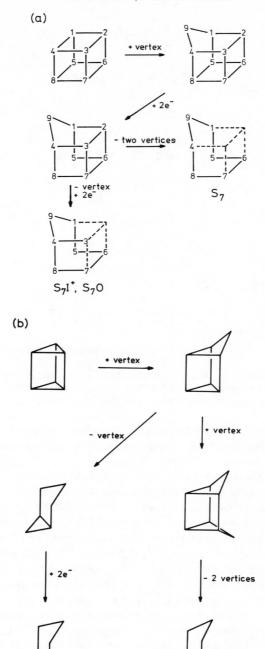

(c)

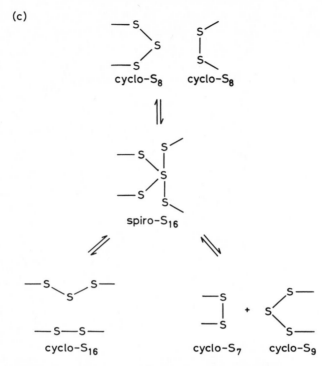

Figure 3.4 (a), (b) Relationship between the structures of some sulphur rings and parent 'electron-precise' clusters; (c) possible pathway for the conversion of S_8 into S_7 and S_9.

to be favoured and the rings therefore consist of chains of sulphur atoms attached to chains of selenium atoms, in contrast to the *cyclo*-sulphur imides (p. 49) where S—N bonds are favoured over N—N bonds.

Cyclo-sulphur rings are yellow or pale orange, low-melting solids of variable stability. For example, S_9 decomposes above $0\,^{\circ}$C whereas S_6 and S_7 are only moderately stable at room temperature. Their interconversion reactions ($S_n \longrightarrow S_m$; $m \neq n$) have been studied in some detail.[2] Both thermal and photochemical reactions are possible and a number of mechanisms have been proposed: (1) free radical, consisting of ring opening ($S_n^{\text{ring}} \longrightarrow S_n^{\text{chain}}$), polymerization ($S_n^{\text{chain}} \longrightarrow S_\infty^{\text{chain}}$) and depolymerization ($S_\infty^{\text{chain}} \longrightarrow$ mixture of $S_m^{\text{ring}} + S_n^{\text{ring}}$); (2) isomerization/transfer which involves formation of an intermediate ($S_8 \longrightarrow S_7S$) with a similar structure to S_7O followed by sulphur transfer ($S_7S + S_8 \longrightarrow S_7 + S_9$); (3) dimerization followed by dissociation (Fig. 3.4c). It seems likely that the last pathway is important at moderate temperatures ($\leqslant 150\,^{\circ}$C), whilst there is evidence for free radical intermediates in photochemical reactions and at higher temperatures.

Cyclo-S_6 is known[7] to be more susceptible to nucleophilic attack than *cyclo*-S_8, and this feature may be of utility in organic synthesis. A fascinating coordination compound of S_8 has been reported:[8]

$$AgAsF_6 + 2S_8 \xrightarrow{\text{liq. SO}_2} [Ag(S_8)_2][AsF_6]$$

There are very few reactions known for S_nO species, although adducts with Lewis acids ($SnCl_4$, $SbCl_5$) have been prepared.[9,10] The structure of $[SnCl_4(S_8O)_2]$ is a distorted octahedral arrangement of four chlorine and two oxygen coordinated *cis*-S_8O ligands.[11]

On p. 43 we mentioned the use of Cp_2TiS_5 as a synthetic reagent in the preparation of *cyclo*-sulphur species. It is worth noting that there are a wide variety of metal-substituted *cyclo*-sulphur systems with various ring sizes. This area was recently reviewed.[12] Dianionic ligands, $S_n{}^{2-}$ ($n = 3$–9), have been stabilized by complexation. Particularly interesting are $Pd_2S_{28}{}^{4-}$, which is a binuclear complex with four bridging $S_7{}^{2-}$ ligands,[13] and $Bi_2S_{34}{}^{4-}$, in which the two bismuth atoms are coordinated by two $S_7{}^{2-}$ ligands and linked by an $S_6{}^{2-}$ group.[14] A silver complex containing the $S_9{}^{2-}$ ligand is known[15] and a cubane-like cluster $Re_4S_{22}{}^{4-}$ was recently reported.[16]

3.2 CYCLIC SULPHUR IMIDES

The NH or NR group is isoelectronic with an S atom and it might be expected that sulphur–nitrogen rings based on the cyclic S_n rings described above should exist. In fact, the chemistry of *cyclo*-sulphur imides dates back to 1908 when $S_4(NH)_4$ was first synthesized, and is fairly extensive with compounds of the general formula $S_{8-n}(NH)_n$ ($n = 1$–4) being known.[17] Tetrasulphur tetraimide is prepared by reduction of tetrasulphur tetranitride in methanol. The other compounds are obtained from the reaction of dichlorosulphane with ammonia in an inert solvent such as dimethylformamide.

$$S_4N_4 + SnCl_2 \xrightarrow{\text{MeOH}} S_4(NH)_4$$

$$S_2Cl_2 + NH_3(g) \longrightarrow S_7(NH) + S_6(NH)_2 \text{ (3 isomers)} + S_5(NH)_3$$

More recently a very useful synthesis of $S_7(NH)$ by acid hydrolysis of the S_4N^- anion (prepared in hexamethylphosphoramide) has been reported.[18] Finally, the synthesis of some six-membered imides $[S_4(NR)_2, R = Et, CH_2Ph]$ using high-dilution cyclization has been described.[19]

$$S_8 + NaN_3 \xrightarrow{\text{HMPA}} S_4N^- \xrightarrow{\text{HCl}} S_7(NH)$$

$$40\% \text{ yield}$$

$$2S_2Cl_2 + 6RNH_2 \longrightarrow S_4(NR)_2 + 4RNH_3Cl$$

Selenium imides, although not very common, are also known. Eight- and fifteen-membered rings have been prepared using the lithium salt of (trimethylsilyl)-*tert*-butylamine; the products are stabilized by the bulky ^{t}Bu substituents:[20]

$$(Me_3Si)(^tBu)NLi + SeOCl_2 \longrightarrow 1,4\text{-}[Se_6(N^tBu)_2]$$

$$+ 1,3,6,8,11,13\text{-}[Se_9(N^tBu)_6]$$

$$+ LiCl + (Me_3Si)_2O$$

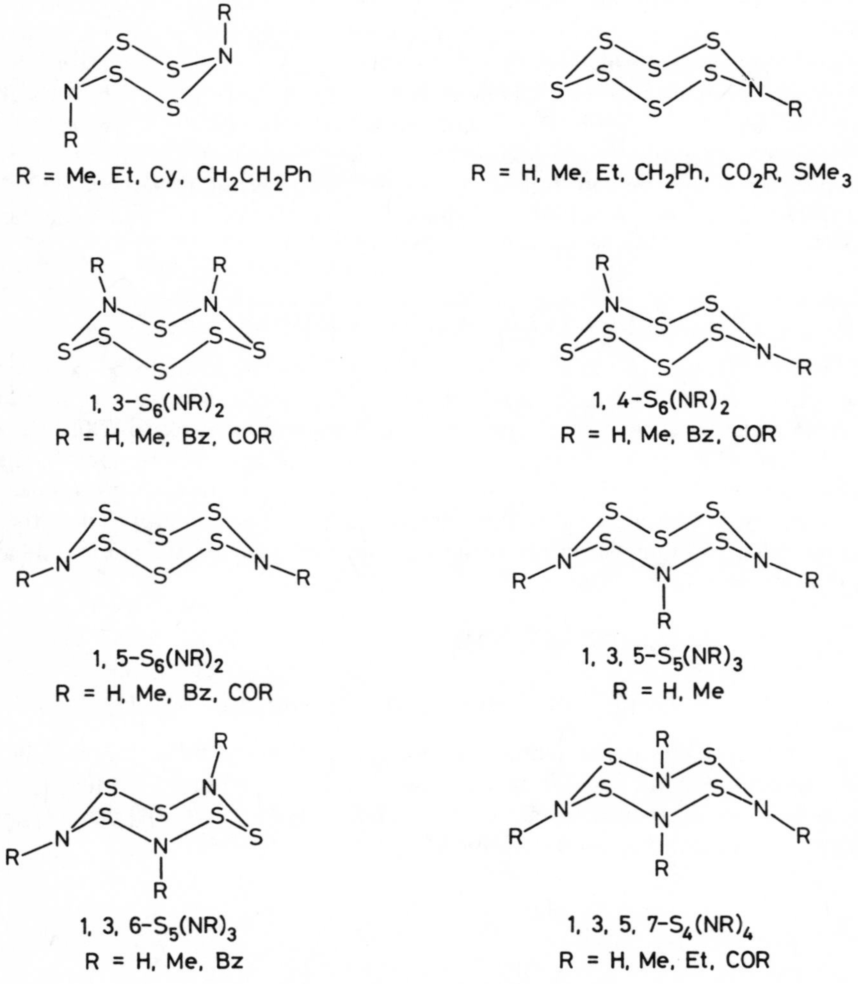

R = Me, Et, Cy, CH_2CH_2Ph

R = H, Me, Et, CH_2Ph, CO_2R, SMe_3

1, 3-$S_6(NR)_2$
R = H, Me, Bz, COR

1, 4-$S_6(NR)_2$
R = H, Me, Bz, COR

1, 5-$S_6(NR)_2$
R = H, Me, Bz, COR

1, 3, 5-$S_5(NR)_3$
R = H, Me

1, 3, 6-$S_5(NR)_3$
R = H, Me, Bz

1, 3, 5, 7-$S_4(NR)_4$
R = H, Me, Et, COR

Figure 3.5 Structures of some cyclic sulphur imides.

The *cyclo*-sulphur imides are colourless, air-stable solids with structures which are based on their isoelectronic sulphur analogues. The eight-membered rings have crown geometries while the six-membered compounds have chair cyclohexane conformations (Fig. 3.5). As can be seen, some of the compounds can exist as isomers with NH substitution occurring at different sites, but none of the crystallographically characterized examples contain neighbouring nitrogen atoms. The synthetic methods are not likely to produce isomers of this type and, if they were formed, elimination of dinitrogen would be energetically very favoured.

In the eight-membered rings the nitrogen atoms are close to trigonal and can be considered to be sp^2 hybridized. This leads to S—N bond lengths (*ca* 1.67 Å) which are slightly shorter than 'pure' S—N single bonds and means that the nitrogens do not function as Lewis bases (see below). Interestingly, in the six-membered rings the geometry around the nitrogen atoms is close to tetrahedral with longer S—N bonds (1.72 Å) and, surprisingly, axial substitution. In cyclohexanes steric interactions are reduced when the substituents are equatorial. Here, electron repulsions between the nitrogen and sulphur lone pairs may be important, and these are minimized when the nitrogen lone pairs are equatorial.

The structures of the two known selenium imides are shown in Fig. 3.6. The eight-membered ring compound has a similar structure to $S_6(NH)_2$; there is no sulphur–nitrogen or *cyclo*-sulphur analogue of the fifteen-membered ring.

Illustrative reactions of $S_7(NH)$ are shown in Fig. 3.7. Most of the reactions are simple substitutions of the NH group and are readily understood. We have

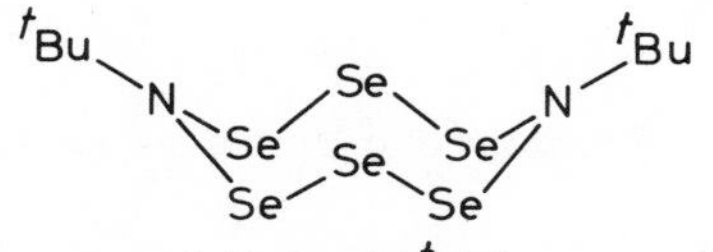

1, 5-$Se_6(N^tBu)_2$

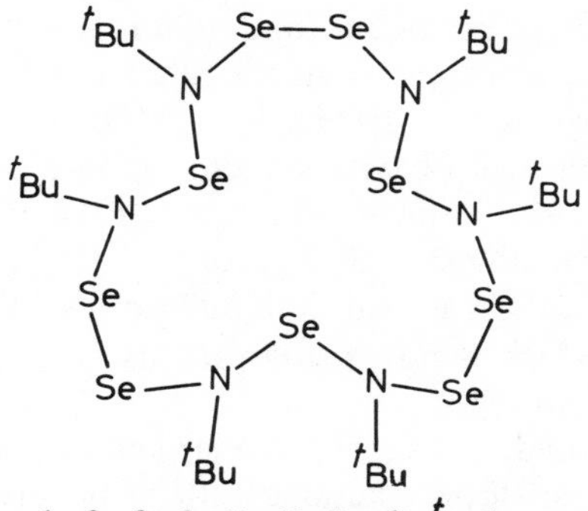

1, 3, 6, 8, 11, 13-$Se_9(N^tBu)_6$

Figure 3.6 Cyclic selenium imides.

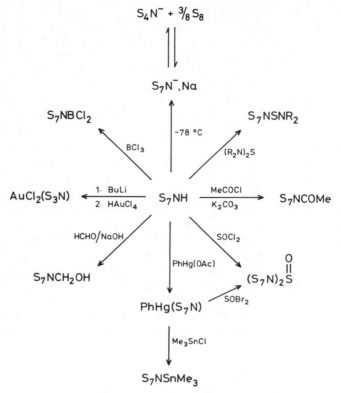

Figure 3.7 Reactions of cyclo-heptasulphur monoimide.

already alluded to the trigonal nature of the nitrogen atom, and this is reflected in the condensation reactions with Lewis acids, e.g. BCl_3. Coupling reactions of $S_7(NH)$, e.g. using $SOCl_2$, provide a route to larger structures. Fused rings can also be formed when an appropriate starting material is chosen. Reaction of $1,3$-$S_6(NH)_2$ with S_5Cl_2 gives $S_{11}N_2$ (Fig. 3.8), which consists of two fused eight-membered crown rings again having trigonal geometry at the nitrogen atoms.

The reactions of *cyclo*-sulphur imides with bases are perhaps mechanistically the most important. Treatment of $S_7(NH)$ or $S_4(NH)_4$ with sodium hydroxide at room temperature produces a beautiful, intense inky blue solution due to the linear S_4N^- anion, which is now recognized as an important intermediate in sulphur–nitrogen chemistry.[21]

Tetrasulphur tetraimide undergoes similar substitution reactions and also a number of poorly understood reactions with metal salts (Fig. 3.9). It has been shown to coordinate through sulphur in complexes with Group VI carbonyls, but with platinum complexes cleavage to disulphurdinitrido complexes takes place.

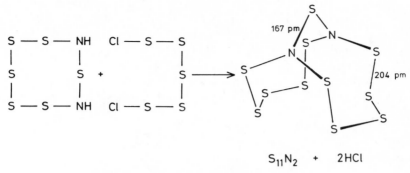

Figure 3.8 Preparation and structure of $S_{11}N_2$.

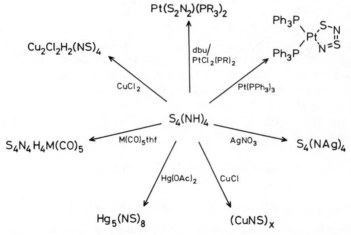

Figure 3.9 Formation of metal–sulphur–nitrogen compounds from tetrasulphur tetraimide.

3.3 CYCLOPOLYPHOSPHANES AND PHOSPHIDES

3.3.1 Preparation and Structures

Although the PH group is isoelectronic with the sulphur atom, the variety and complexity of structures adopted are much greater for the phosphorus compounds.[22] Like boranes, the simple phosphanes can be divided into a several homologous series stretching from P_3H_5 to $P_{18}H_4$, although many of the compounds have only been observed by mass spectrometry or ^{31}P NMR.

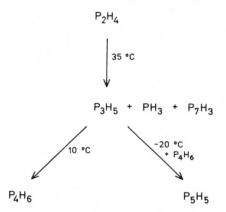

Figure 3.10 Use of P_2H_4 in the preparation of higher phosphanes.

Currently, only P_2H_4, P_3H_5, P_4H_7 and P_7H_7 can be isolated as pure compounds. As predicted from empirical bonding descriptions, the P_nH_{n+2} species (isoelectronic with S_n^{2-}) are open chains; P_nH_n (isoelectronic with S_n) are simple rings, and the more phosphorus-rich compounds are cages. Apart from the hydrides, a substantial number of P_nR_n rings are known.[23]

Most of the phosphanes are obtained from thermolysis reactions of diphosphane (Fig. 3.10), although methanolysis of silyl derivatives provides a more attractive route to P_5H_5 and P_7H_3:

$$(Me_3SiP)_4 \xrightarrow{\text{MeOH}} P_5H_5 + Me_3SiOMe$$

$$(Me_3Si)_3P_7 \xrightarrow{\text{MeOH}} P_7H_3 + 3Me_3SiOMe$$

Organophosphanes have been known for many years and may be obtained by a variety of routes. Dehalogenation, or reaction with RPH_2, of dichloroorganophosphines gives mixtures of different ring sizes:

$$RPCl_2 + Mg \longrightarrow 1/n(RP)_n + MgCl_2$$

$$RPCl_2 + RPH_2 \longrightarrow 2/n(RP)_n + 2HCl$$

Three-membered rings are readily obtained from $[2 + 1]$ cyclocondensation reactions:

$$RPCl_2 \quad + \quad \begin{matrix} R & & R \\ \diagdown & & \diagup \\ & P\text{—}P & \\ \diagup & & \diagdown \\ H & & H \end{matrix} \quad \longrightarrow \quad (RP)_3 \quad + \quad 2HCl$$

Larger, more phosphorus-rich cage/bicyclic compounds can be prepared[22] by the reactions shown below:

$$3MePCl_2 + 6PCl_3 \xrightarrow{Mg} P_9Me_3 + 9MgCl_2 + 3Cl_2$$

varying the proportions of reagents used allows isolation of most of the known $P_{x+y}R_x$ species. Related arsenic compounds are obtained from similar reactions.[24] A simple rational route (relying on the formation of Me_3SnCl) to 1, 1-bicyclotriphosphine was recently described:[25]

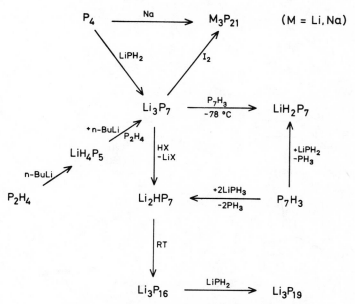

Phosphides are obtained as shown in Fig. 3.11; mixed $[P_{7-x}As_x]^{3-}$ species also exist.

Figure 3.11 Preparative routes to phosphides.

Illustrative structures of *cyclo*-phosphanes are shown in Fig. 3.12. The solid-state structures of P_nR_n may be different to the situation in solution. For example, the five-membered ring system is, according to X-ray crystallography, distorted from planar to different extents depending on the R group. However, ^{31}P NMR studies are in accord with a planar structure, although the observed AA'BB'C spectra can be pictured as resulting from a rapid puckering of the ring, which gives time-averaged 2:2:1 phosphorus environments. Clearly, the energy difference between planar and puckered geometries is small. Six-membered ring compounds have chair cyclohexane structures with the R groups equatorial. The P—P bond distances in the rings are *ca* 2.25 Å and their UV spectra have absorptions not observed in trialkylphosphines. This, together with the observ-

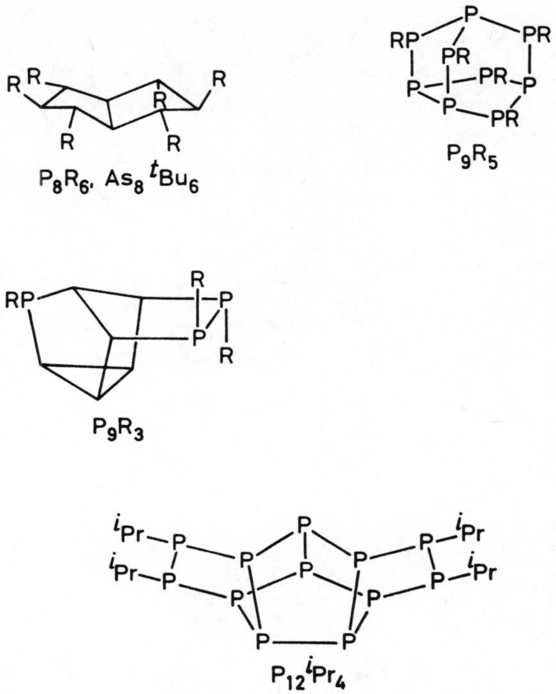

Figure 3.12 Structures of substituted phosphanes and arsanes.

ation that the compounds are non-basic [$pK_a = 1$ for $(PR)_4$] relative to tertiary phosphines, has led to the suggestion that there is some electron delocalization between the lone pair of one phosphorus atom and a vacant d orbital of a neighbouring phosphorus atom.[26]

Many of the large cyclo-phosphanes form cages with complex structures; however, some have structures which are closely related. P_7R_3, P_9R_5 and P_9R_3 have structures which are based on a P_4 tetrahedron. The first two also relate to the isoelectronic phosphorus sulphides P_4S_3 and P_4S_5. The other, more open, compounds can usefully be compared with organic equivalents ($P_7R_5 \equiv$ norbornane) or with other main group rings, and this has been discussed in Chapter 1. The structures of the phosphide anions are shown in Fig. 3.13; in general they are based on P_7^{3-} units which are bridged in larger cages. Interestingly, in the mixed $[P_{7-x}As_x]^{3-}$ systems it is predicted that the bridge atoms should be As^- since arsenic is more electronegative than phosphorus. Although this is the case when the counterion is Li^+ or Na^+, when Rb^+ is used the material interconverts between several isomers.[27]

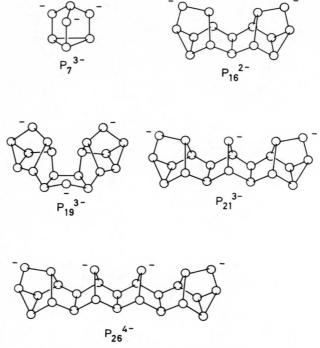

Figure 3.13 Structures of phosphide anions.

3.3.2 Reactions

A wide variety of reactions are known for the organic derivatives, although relatively few have been reported for the parent compounds. Reactions with metal carbonyls often proceed with ring preservation and in these cases the organocyclophosphanes can act as 2-electron donors, e.g. $[Cr(CO)_5\{(P^iPr)_3\}]$ or, by bridging metals, as 4-electron donors, e.g. $[\{Cr(CO)_5\}_2\{(P^iPr)_3\}]$.[22] An alternative route to a triply bridging $(PPh)_3$ has also been published.[28]

Interestingly, reaction[29] of $[Mo(CO)_3Cp]_2$ with $(AsMe)_5$ gives $[Cp(CO)_2Mo(\mu^2 - AsMe)_5Mo(CO)_2Cp]$ or, under more vigorous conditions, $[CpMo(As)_5MoCp]$, a triple-decker sandwich compound containing the As_5 moiety which is isoelectronic with Cp. In fact the stabilization of phosphorus and arsenic rings by metals has become commonplace, with three- to six-membered rings such as P_3^{3-}, P_4, P_5 and P_6 all having been obtained[30,31] (there has even been an analogue of ferrocene with a P_5 ring reported[32]). Figure 3.14 shows some representative compounds.

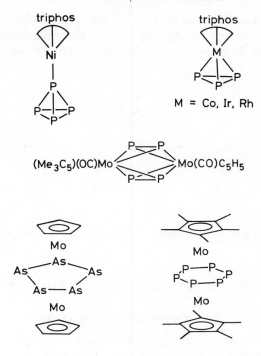

Figure 3.14 Some metal complexes containing unsupported Group V ligands.

Complete degradation of the rings occurs very readily with a wide variety of reagents, e.g. halogens, oxygen, ammonia, water, alcohols and alkali metals. Generally, most compounds are stable if stored in dry, oxygen-free conditions in the cold. Careful addition of dry air to $P_6{}^tBu_4$ or $P_7{}^tBu_5$ gives $P_6{}^tBu_4O$ or $P_7{}^tBu_5O$, respectively; in both compounds ^{31}P NMR suggests that the ring structures are retained with a simple oxidation of one phosphorus having occurred.[33] Hexa-*tert*-butyloctaphosphane is stable to air for several weeks but does react with cumene hydroperoxide:[34]

$$\begin{array}{c} Me \\ | \\ P_8{}^tBu_6 + 6PhCOOH \longrightarrow P_8{}^tBu_6O_6 \\ | \\ Me \end{array}$$

Addition reactions occur; with sulphur[35] (PhP)$_5$ gives (PhP)$_4$S. Treatment of (P^{t}Bu)$_3$ with sulphur gives the monothiocyclotriphosphane which rapidly rearranges at room temperature to a four-membered heterocyclic ring:

Finally, rearrangements between different ring sizes have been extensively studied.[35]

3.4 PHOSPHORUS–OXYGEN/SULPHUR CAGES

3.4.1 Preparation and Structures

The five simple binary oxides and six well established sulphides[36] are shown in Fig. 3.15 together with examples of the mixed P—O—S cages.[37] Closely related to these are P—N cages such as P$_4$(NBu)$_4$[38] and mixed phosphorus–arsenic–suphides (P$_x$As$_{4-x}$S$_3$) and phosphorus–sulphur–selenides, P$_4$S$_x$Se$_{3-x}$.[39] Structurally, all of the compounds can be derived from the P$_4$ tetrahedron by edge bridging or vertex addition of oxygen or sulphur atoms. It is notable that there are only two isostructural oxides and sulphides and thus some structural types only exist as oxides or sulphides (e.g. P$_4$O$_6$ and P$_4$S$_5$). An alternative viewpoint of the structures of the lower phosphorus sulphides (P$_4$S$_4$, P$_4$S$_5$ and P$_4$S$_7$) is useful. For example, in P$_4$S$_4$, looking at the cage with respect to the sulphur atoms shows that these are in a square plane with the phosphorus atoms in a P$_4$ tetrahedron above and below the plane; P$_4$(N^{t}Bu)$_4$ has a similar structure with N^{t}Bu groups replacing sulphur atoms. This viewpoint underlines the structural similarity of P$_4$S$_4$ with Se$_4$N$_4$, P$_4$(N^{t}Bu)$_4$ and S$_4$N$_4$, although in the latter case the Group V and VI atoms are interchanged. The structure of P$_4$S$_5$ is analogous to S$_4$N$_5{}^-$ but there is, as yet, no sulphur–nitrogen analogue of P$_4$S$_7$. Care must be taken in pressing this type of analogy too far; although the gross geometries appear similar, there are significant differences in the bonding descriptions. All of the phosphorus sulphides may be adequately described by simple two-centre, two-electron bonds and there is no indication of additional cluster bonding. The situation for sulphur–nitrogen compounds is more complex and will be discussed later.

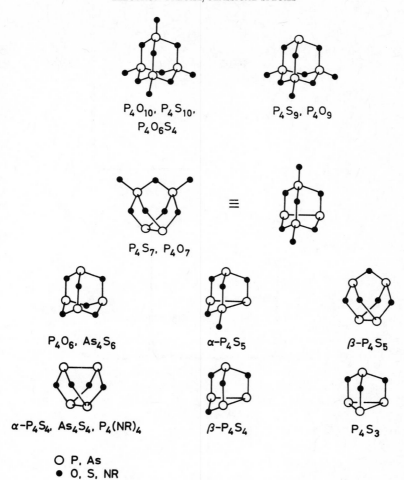

O P, As
• O, S, NR

Figure 3.15 The known phosphorus sulphides, phosphorus oxides and arsenic sulphides.

Whilst considering isoelectronic analogies, we should note that P_4S_3 and P_7^{3-} have very similar structures with P^- being equivalent to S and that a number of $P_4(NR)_x$ compounds also exist.

Originally, phosphorus oxides and sulphides were prepared by direct combination of the elements and for many compounds this is still a satisfactory procedure. Typically, P_4O_{10} is obtained by burning phosphorus in dry air, the vapourised product being condensed out of the reaction. P_4O_6 is prepared by

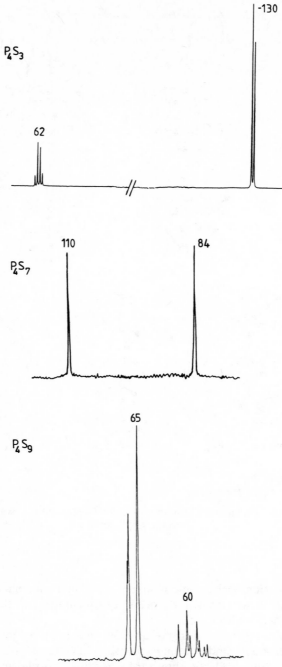

Figure 3.16 ^{31}P NMR spectra of (from top to bottom) P_4S_3, P_4S_7 and P_4S_9.

controlled oxidation of white phosphorus at *ca* 50 °C. The phosphorus sulphides P_4S_3, P_4S_7 and P_4S_{10} are produced commercially by reaction of white phosphorus and sulphur at elevated temperatures (over 300 °C) under an inert atmosphere with the products being distilled from the reactions. Convenient laboratory-scale preparations use red phosphorus and sulphur, often in an inert solvent (e.g. xylene). Combination of P_4S_{10} or P_4S_3 with the appropriate quantity of phosphorus or sulphur can be used to give P_4S_5, P_4S_7 and P_4S_9:

$$P_4S_3 + 1/2S_8 \longrightarrow I_2P_4S_7$$

Some milder and more elegant strategies have been developed that are useful for the thermally unstable compounds P_4S_5 and P_4S_4. Triphenylphosphine can be progressively used as a desulphurizing agent:[40]

$$P_4S_{10} + PPh_3 \longrightarrow P_4S_9 + SPPh_3$$

$$\downarrow 2PPh_3$$

$$\beta\text{-}P_4S_5 \leftarrow P_4S_7 + 2SPPh_3$$

Alternatively, the tin reagent $(Me_3Sn)_2S$ is useful:[41]

$$P_4S_3 + I_2 \longrightarrow \beta\text{-}P_4S_3I_2 \longrightarrow \alpha\text{-}P_4S_3I_2$$

$$\downarrow (Me_3Sn)_2S$$

$$\alpha\text{-}P_4S_4 + 2Me_3SnI$$

The ^{31}P NMR spectra of phosphorus sulphides provide an interesting example of the use of NMR in identifying cage compounds. As expected, the spectrum of P_4S_{10} consists of a singlet whereas that of P_4S_3 contains two resonances—a doublet and a quartet. Since ^{31}P has 100% natural abundance and $I = 1/2$, it is straightforward to assign the two signals to the basal and apical phosphorus atoms, respectively. The spectra of the other cages are more complex (e.g. P_4S_9 gives a very second-order spectrum) but are useful for identification purposes (Fig. 3.16).

3.4.2 Reactions

With the exception of P_4S_3, the phosphorus sulphides and oxides are sensitive to moisture and readily hydrolysable. The relative stability of P_4S_3 is sometimes used in its purification; the impure material is suspended in boiling water, which hydrolyses the other compounds present:

$$P_4O_{10} + 6H_2O \longrightarrow 4H_3PO_4$$

$$P_4S_{10} + 16H_2O \longrightarrow 4H_3PO_4 \rightarrow 10H_2S$$

$$P_4O_6 + 6H_2O \longrightarrow 4H_3PO_3$$

Reactions with alcohols lead to mono- and dialkoxy acids:

$$P_4O_{10} + 6ROH \longrightarrow 2(RO)PO(OH)_2 + 2(RO)_2PO(OH)$$

$$P_4S_{10} + 8ROH \longrightarrow 4(RO)_2PS(SH) + 2H_2S$$

The dithiophosphosphoric acids are very important commercially for the preparation of pesticides such as malathion and in the manufacture of engine oil additives such as zinc dialkyldithiophosphates (ZDDPs), which have interesting cluster geometries.[42]

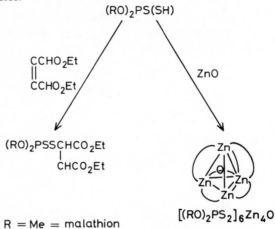

$R = Me = $ malathion

All of the phosphorus sulphides burn readily in air and P_4S_3 is still widely used (in combination with $KClO_4$ and binders) in 'strike anywhere' matches.

We have already mentioned the reaction of iodine with P_4S_3; this compound and the others which contain P—P bonds (P_4S_5 and P_4S_7) also react with bromine. The first stage of the reaction may be an opening of the P—P bond:

$$P_4S_7 + 3Br_2 \longrightarrow P_2S_6Br_2 + P_2S_5Br_4$$

Reactions with chlorine completely destroy the cages; P_4S_{10} gives S_2Cl_2, SCl_2 and PCl_5.

Reaction of P_4S_{10} with KCN/H_2S yields the trithiometaphosphate anion, PS_3^-, which has been stabilized using the bulky cation Ph_4As^+ to stop dimerization to $P_2S_6^{2-}$. The latter can be prepared directly by heating together an alkali metal with the appropriate quantities of phosphorus and sulphur. With NaCN (but no H_2S) $[(NCPS_2)_2S]^{2-}$ is obtained:

$$P_4S_{10} \xrightarrow{\text{NaCN}} \begin{bmatrix} & S & & S & \\ & | & & \| & \\ NC{-}P{-}S{-}P{-}CN \\ & \| & & | & \\ & S & & S & \end{bmatrix}^{2-}$$

Treatment of P_4S_{10} with NaN_3 gives $P(N_3)_2S_2{}^-$; this reacts with additional P_4S_{10} to give $[P_4S_9N]^-$ in which an N^- has replaced a bridging sulphur atom.[43]

A number of metal complexes in which P_4S_3 acts as a donor ligand through its apical phosphorus have been prepared:

$$Ni(\eta^3\text{-}C_3H_5)_2 + 4P_4S_3 \longrightarrow Ni(P_4S_3)_4$$

$$Mo(CO)_6 + P_4S_3 \longrightarrow Mo(CO)_5(P_4S_3)$$

Insertion into the basal P—P bond is also possible. For example, the oxidative addition reaction of P_4S_3 to Vaska's compound, trans-Ir(CO)Cl(PPh$_3$)$_2$ in benzene at $60\,^\circ C$ gives an octahedral Ir^{III} complex, $[Ir(CO)Cl(P_4S_3)(PPh_3)]_2$ (Fig. 3.17), as a green, air-stable solid. Related reactions have been described.[30,31]

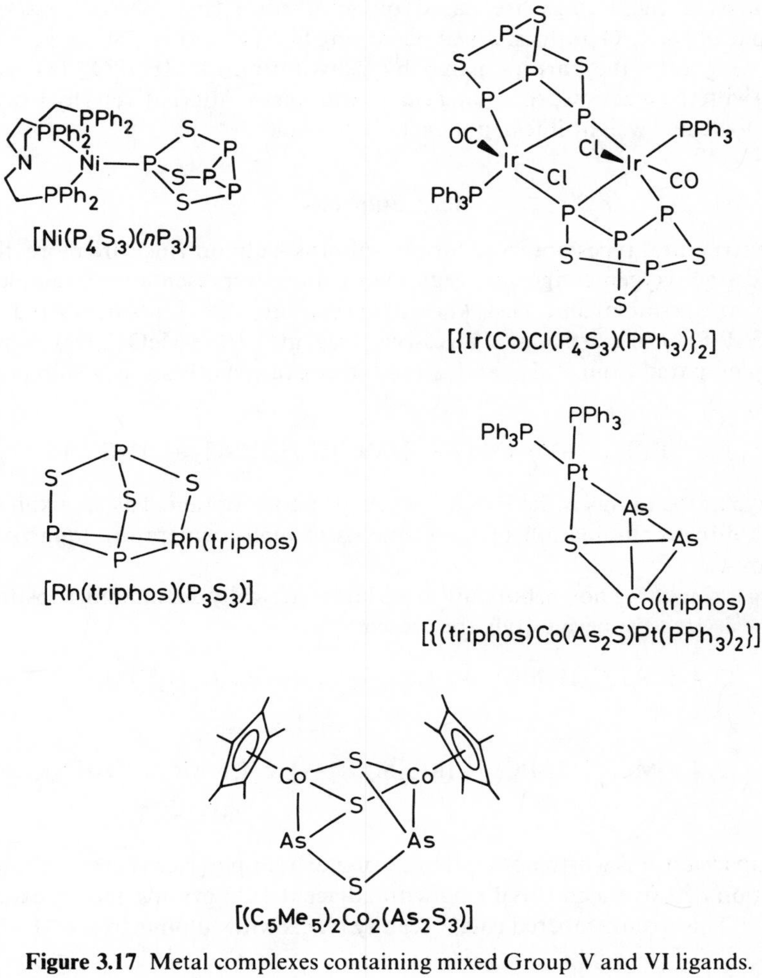

[Ni(P$_4$S$_3$)(nP$_3$)]

[{Ir(Co)Cl(P$_4$S$_3$)(PPh$_3$)}$_2$]

[Rh(triphos)(P$_3$S$_3$)]

[{(triphos)Co(As$_2$S)Pt(PPh$_3$)$_2$}]$^+$

[(C$_5$Me$_5$)$_2$Co$_2$(As$_2$S$_3$)]

Figure 3.17 Metal complexes containing mixed Group V and VI ligands.

Metal centres can also be used to trap reactive arsenic–sulphur species (Fig. 3.17):

$$[C_5Me_5)Co(CO)_2]_2 + As_4S_4 \xrightarrow[hv]{thf} [(C_5Me_5)_2Co_2(As_2S_3)]$$

Reaction of P_4S_{10} with anisole gives Lawessons reagent,[44] discussed below.

3.5 PHOSPHORUS–OXYGEN/SULPHUR RINGS

3.5.1 Oxides

Cyclic phosphates are well known,[45] in particular the 'so-called' metaphosphates in which the parent acid has the formula $(HPO_3)_n$ ($n = 3, 4, 5, 6, \ldots$). The structures of these rings are based on tetrahedral (sp^3) phosphorus(V); for example, in $Na_3(PO_3)_3$ the six-membered ring has a chair cyclohexane conformation. Generally they are prepared by dehydration of MH_2PO_4 (M = alkali metal) with the reaction proceeding via several stages. Alternatively, hydrolysis of high molecular weight linear polymers is possible.

3.5.2 Sulphides

The structural types observed for phosphorus–sulphur rings are more diverse than for the oxygen congeners. Figure 3.18 shows representative examples.

The most stable and best known species are the four-membered rings $RP(S)S_2P(S)R$, in particular Lawessons reagent[44] (R = $MeOC_6H_4$), which is readily prepared from P_4S_{10} and is used in organic synthesis as a sulphurating agent:

$$P_4S_{10} + 4PhOMe \longrightarrow 2[(MeOC_6H_4)PS_2]_2 + 2H_2S$$

Closely related to this is the $P_2S_6{}^{2-}$ anion, generally obtained as an alkali metal salt by direct combination of the elements in high-temperature (quartz tube) reactions.

Larger rings are known, but only when large, sterically demanding substituents (e.g. mesityl and supermesityl) are present:

$$(2, 4, 6\text{-}^tBu_3C_6H_2)PCl_2 + Li_2S \longrightarrow [(2, 4, 6\text{-}^tBu_3C_6H_2)PS]_3$$
$$+ 2LiCl$$
$$(2, 4, 6\text{-}Me_3C_6H_2)PCl_2 + (Me_3Si)_2S \longrightarrow [(2, 4, 6\text{-}Me_3C_6H_2)PS]_4$$
$$+ 2Me_3SiCl$$

In comparison, it is worth noting that the use of a simple phenyl group allows the formation of a five-membered ring with adjacent PPh groups and an exocyclic sulphur. This five-membered ring[46] readily reacts with sulphur to give $(PhPS_2)_2$:

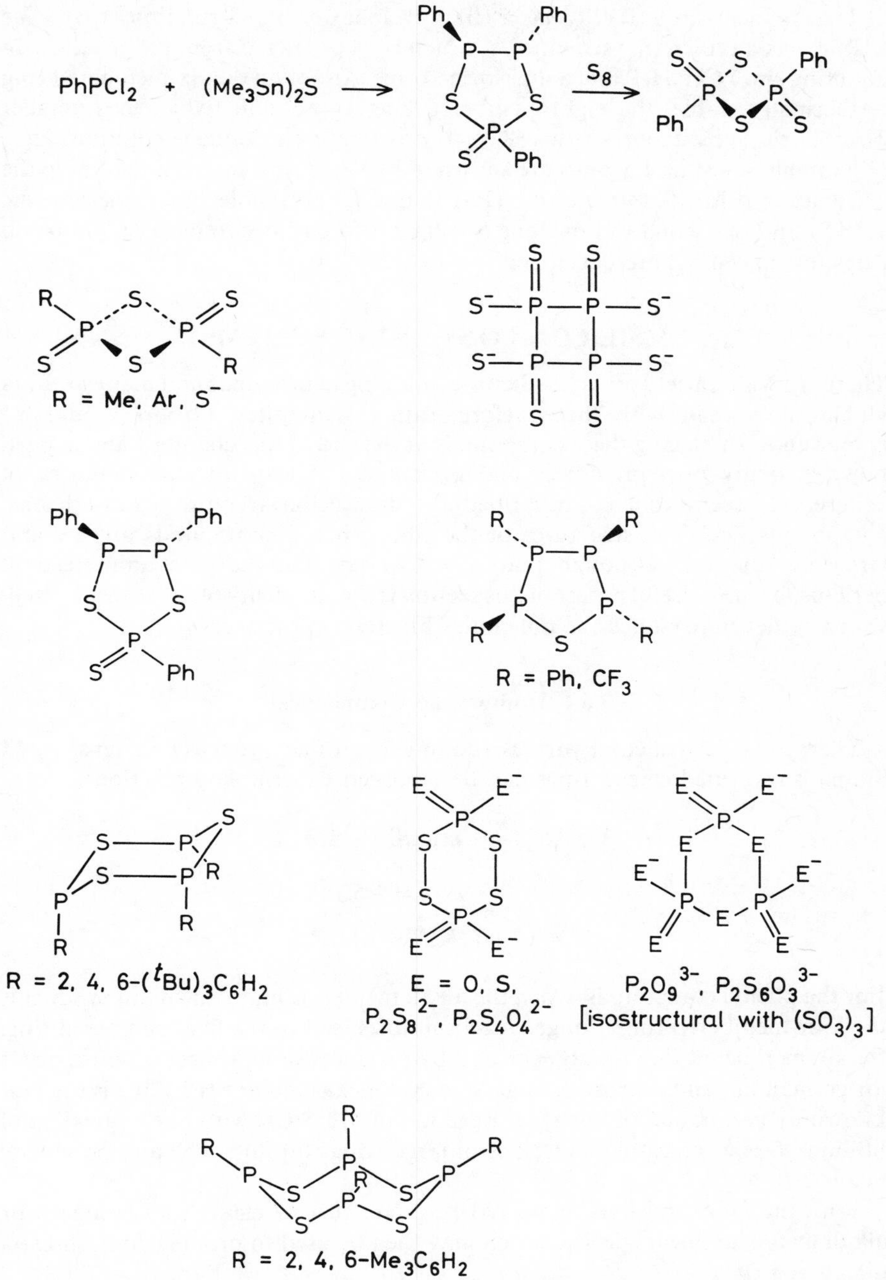

Figure 3.18 Cyclic phosphorus sulphides and oxides.

The six-membered $(RPS)_3$ has a chair cyclohexane conformation,[47] whereas $(RPS)_4$ has a crown (S_8) structure,[48] which is as predicted from simple electron counting since CH_2, RP and S are isoelectronic. Although the six-membered ring is thermally stable, the eight-membered ring compound (which has smaller phosphorus substituents) forms $(RPS_2)_2$ on standing in toluene solution.

Examples of other PS rings are shown in Fig. 3.18. It is interesting to note the existence of $(PhP)_4S$, formed from $(PhP)_5$ and S_8. This molecule is isoelectronic with S_5 and one wonders how long it will be before routes to more sulphur-rich rings (or indeed S_5) are developed.

3.6 SILICON-CONTAINING SYSTEMS

There is a wide variety of cyclic silicon-containing compounds and many authors would have dealt with these before, rather than after, Group V and VI compounds. In placing these compounds at the end of this chapter I am, in part, bowing to my own prejudices and preferences. The main reason is one of coherence; it seems to me to be far better to discuss the structurally most diverse compounds first. The structures of the silicon-based compounds are, by and large, not unique—although that is not to say that they are uninteresting! perhaps the most useful aspect of this section is the simplicity of the reagents used, which rather understates the difficulties inherent in the area.

3.6.1 Homocyclic Compounds

There are many silicon hydrides and in general they are fairly air sensitive.[49] Simple substituted silicon rings can be prepared via coupling reactions:

$$R_2SiCl_2 \xrightarrow{2M} 1/n(R_2Si)_n + 2MCl$$

$$(n = 4, 5, 6)$$

$$(M = Li, Na, K; R = Me, Ph)$$

For the phenyl compounds, when the alkali metal is lithium, the main product is the four-membered ring. Longer reaction times lead to the five-membered ring; the six-membered ring is only seen as a by-product. Similar reactions also occur for germanium and tin; interestingly, the cyclohexastannane $(Ph_2Sn)_6$ is the best known tin compound of this type. Reaction of Me_2SiCl_2 with one equivalent of lithium gives appreciable quantities of larger rings (up to $n = 9$) and polymeric material.

Both the four- and five-membered rings are readily cleaved by halogens or alkali metals to linear species, which may then be used to produce five- and six-membered rings:

$$(Ph_2Si)_4 + 2Li \longrightarrow Li(Ph_2Si)_4Li$$

$$\downarrow \text{Me}_2\text{SiCl}_2$$

$$(Ph_2Si)_4(Me_2Si)_n$$

$$(n = 1, 2)$$

or to introduce heteroatoms such as platinum:

$$(Ph_2Si)_5 + Br_2 \longrightarrow Br(Ph_2Si)_5Br \xrightarrow{\text{LiAlH}_4} H(Ph_2Si)_5H$$

$$\downarrow \text{Pt(C}_2\text{H}_4)(\text{PPh}_3)_2$$

$$(Ph_2Si)_4Pt(PPh_3)_2$$

There are also reactions in which the phenyl groups may be displaced. For example, treatment of $(Ph_2Si)_n$ ($n = 5, 6$) with HI or HBr substitutes phenyl by halogen, with the extent of substitution being dependent on n and the halogen; for $n = 5$ and HBr the decabromo compound is obtained, but HI is less reactive as is $n = 6$, which gives the hexabromo species with HBr. It is assumed from NMR studies on $(PhSiH)_5$ [obtained from $(PhSiBr)_5$ and $LiAlH_4$] that in partially substituted compounds each silicon has one halogen/hydrogen and one Ph group.

The structures of *cyclo*-polysilanes are as might be expected; the six-membered rings adopt chair cyclohexane-like geometries, the five-membered rings are puckered rings [like the isoelectronic $(PR)_5$ system], whereas the four-membered rings have butterfly structures. An interesting example of a four-membered ring with a cross-ring bond was reported recently.[50] $(Ar_2Si)_2({}^tBuSi)_2$ [Ar = 2, 6-$(C_2H_5)C_6H_3$-], obtained by reduction of $(Ar_2Si)({}^tBuSiCl)_2$, exists as a butterfly structure with a cross-ring Si—Si bond and is isostructural with *arachno*-B_4H_{10}, as predicted from simple electron counting rules. The Si—Si bond is cleaved by water.

3.6.2 Silicon–Oxygen Compounds

There are many soluble silicates[51] and a wealth of information on silicate minerals,[52] and we shall not discuss either of these types of compounds here. The chemistry of silicon–oxygen chain compounds (silicones) is very well developed since these types of materials are useful commercially as sealants, etc. A number of cyclic species are known and we shall limit ourselves to a brief discussion of these; the structures of solid silicates are outside the scope of this book.

Cyclic silicones are intermediates in the formation of polymers:

$$R_2SiCl_2 + H_2O \rightarrow 1/n[R_2SiO]_n \xrightarrow{ca\,300\,°C} \text{long chain polymers}$$

$$(n = 3, 4, 5)$$

Structurally, they consist of simple alternating Si—O rings. Since R_2Si and O are both isoelectronic with CH_2 and S, it is not surprising that the rings are similar in geometry to well known species; i.e. for $n = 3$, chair cyclohexane and for $n = 4$, crown S_8.

Polycyclic and spiro compounds are also known and these are generally obtained from thermolysis reactions. Interesting examples which again reflects the usefulness of simple isoelectronic analogies are $(Si^tBu)_4O_6$ and $(HSi)_4S_6$ which have similar structures to P_4O_6 (Si^tBu and SiH are isoelectronic with P). Structurally related to these latter compounds are tin-sulphur, germanium-sulphur and germanium-selenium species, $(RSn)_4S_6$, $(RGe)_4S_6$ and $(RGe)_4Se_6$.[53,54]

$$(H_3Si)_2E + HSiCl_3 \xrightarrow[\text{AlCl}_3]{20\,^\circ\text{C}} H_3SiCl + [(HSi)_4E_6]$$

$$(E = S, Se)$$

3.6.3 Silicon–Nitrogen Compounds[35]

Reaction of gaseous ammonia with R_2SiCl_2 gives $[(R_2Si)NH]_n$ ($n = 3, 4$, R = Me, Et; $n = 3$, R = Ph), whereas with extremely bulky R groups stable silyldiamines, $R_2Si(NH_2)_2$, are formed, but these can be converted to cyclosilazanes on heating. Nitrogen-substituted species are also known and can be obtained from reaction of RNH_2 with R'_2SiCl_2 [e.g. when R = Et and R' = Me, reaction in a sealed tube for 2 weeks gives $(R'_2SiNR)_3$ in 55% yield]:

$$R_2SiCl_2 + 3NH_3 \longrightarrow 1/n(R_2SiNH)_n + 2NH_4Cl$$

$$(n = 3, 4)$$

Generally, the mixtures of $n = 3$ and 4 species can be separated by distillation, but interconversion between the two ring sizes is relatively straightforward with acidic conditions favouring ring expansion and heat causing ring contraction:

$$4(Me_2SiNH)_3 \xrightleftharpoons[400\,^\circ\text{C}]{H_2SO_4} 3(Me_2SiNH)_4$$

Structurally we would expect the silazanes to have familiar ring conformations. For the $n = 3$ case, a flattened chair cyclohexane structure is observed. Interestingly, in the crystal structure of $(Me_2SiNH)_4$, a 1:1 mixture of chair and cradle conformations is adopted, emphasizing the small energy differences between different structures. A large number of four-membered rings are known, with fluorine or organic substituents. A selection of routes is shown below. For a more detailed account of this and the preparation of spiro compounds, see ref. 35.

$$R_2Si(NLiR')_2 \ + \ R''_2SiCl_2 \ \longrightarrow \ R_2Si \underset{\displaystyle \underset{R'}{|}}{\overset{\displaystyle \overset{R'}{|}}{\underset{N}{\overset{N}{\diagup\diagdown}}}} SiR''_2$$

$$(Me_3Si)_2NSiF_2NH^tBu \ \xrightarrow{\ ^nBuLi\ } \ (Me_3Si)_2FSi \underset{\underset{tBu}{|}}{\overset{\overset{tBu}{|}}{\underset{N}{\overset{N}{\diagup\diagdown}}}} SiF(Me_3Si)_2$$

$$2(Me_3SiNSiMe_2)_3 \xrightarrow{\Delta} 3(Me_3SiNSiMe_2)_2$$

3.7 REFERENCES

1. J. Donohue, *The Structures of the Elements*, Wiley, New York, 1974.
2. R. Steudel, *Top. Curr. Chem.*, 1982, **102**, 149.
3. R. Steudel, in *Gmelin Handbuck der Anorganische Chemie, 8 Aufl., Schwefel*, Erganzungasband 3, Springer, Berlin, 1980.
4. R. Steudel and R. Laitinen, *Top. Curr. Chem.*, 1982, **102**, 177.
5. R. Steudel, M. Papavassiliov, E. M. Strauss and R. Laitinen, *Angew. Chem., Int. Ed. Engl.*, 1986, **25**, 99.
6. R. J. Gillespie, *Chem. Soc. Rev.*, 1979, 315.
7. F. Feher and D. Kurz, *Z. Naturforsch., Teil B*, 1969, **24**, 1089.
8. H. W. Roesky, M. Thomas, S. Schimkowiak, W. Pinkert and G. M. Sheldrick, *J. Chem. Soc., Chem. Commun.*, 1982, 895.
9. R. Steudel, T. Sandow and J. Steidel, *J. Chem. Soc., Chem. Commun.*, 1980, 180.
10. R. Steudel, J. Steidel and J. Pickardt, *Angew. Chem., Int. Ed. Engl.*, 1980, **19**, 325.
11. R. Steudel, J. Steidel and T. Sandow, *Z. Naturforsch., Teil B*, 1986, **41**, 951.
12. M. Draganjac and T. B. Rauchfuss, *Angew. Chem., Int. Ed. Engl.*, 1985, **24**, 742.
13. A. Muller, K. Schmitz, E. Krickemeyer, M. Penk and H. Bugge, *Angew. Chem., Int. Ed. Engl.*, 1986, **25**, 453.
14. A. Muller, M. Zimmermann and H. Bugge, *Angew. Chem., Int. Ed. Engl.*, 1986, **25**, 273.
15. A. Muller, M. Romer, H. Bugge, E. Krickemeyer and M. Zimmermann, *Z. Anorg. Allg. Chem.*, 1986, **534**, 69.
16. A. Muller, E. Krickemeyer and H. Bugge, *Angew. Chem., Int. Ed. Engl.*, 1986, **25**, 272.
17. H. G. Heal, *The Inorganic Heterocyclic Chemistry of Sulfur, Nitrogen and Phosphorus*, Academic Press, London, 1980.
18. J. Bojes and T. Chivers, *J. Chem. Soc., Dalton Trans.*, 1975, 1715.

19. R. Jones, D. J. Williams and J. D. Woollins, *Angew. Chem., Int. Ed. Engl.*, 1985, **24**, 760.
20. H. W. Roesky, K. L. Weber and J. W. Bats, *Chem. Ber.*, 1984, **117**, 2686.
21. T. Chivers and R. T. Oakley, *Top. Curr. Chem.*, 1982, **102**, 117.
22. M. Baudler, *Angew. Chem., Int. Ed. Engl.*, 1982, **21**, 492; 1987, **26**, 419.
23. J. L. Mills, in *Homoatomic Rings, Chains and Macromolecules of Main Group Elements* (Ed. A. L. Rheingold), Elsevier, Amsterdam, 1977.
24. M. Baudler and co-workers, *Angew. Chem., Int. Ed. Engl.*, 1984, **23**, 379; 1980, **20**, 123 and 406.
25. M. Baudler and B. Makowa, *Angew. Chem., Int. Ed. Engl.*, 1984, **23**, 987.
26. W. A. Henderson, M. Epstein and F. S. Seichter, *J. Am. Chem. Soc.*, 1963, **85**, 2462.
27. W. Honle and H. G. von Schnering, *Angew. Chem., Int. Ed. Engl.*, 1986, **25**, 352.
28. G. Huttner, H. D. Muller, A. Frank and H. Lorenz, *Angew. Chem., Int. Ed. Engl.*, 1975, **14**, 572.
29. A. L. Rheingold, M. J. Foley and P. J. Sullivan, *J. Am. Chem. Soc.*, 1982, **104**, 4727.
30. P. T. Wood and J. D. Woollins, *Transition Met. Chem.*, 1986, **9**, 358.
31. M. DiVaira and L. Sacconi, *Angew. Chem., Int. Ed. Engl.*, 1982, **21**, 330.
32. O. J. Scherer *et al.*, *Angew. Chem., Int. Ed. Engl.*, 1987, **26**, 59; 1986, **25**, 363.
33. M. Baudler, M. Michels, M. Pieroth and J. Hahn, *Angew. Chem., Int. Ed. Engl.*, 1986, **25**, 471.
34. M. Baudler and J. Germeshausen, *Angew. Chem., Int. Ed. Engl.*, 1987, **26**, 348.
35. I. Haiduc, *The Chemistry of Inorganic Ring Systems, Part 1*, Wiley–Interscience, New York, 1970.
36. H. Hoffman and M. Becke-Goehring, *Top. Phosphorus Chem.* 1976, **8**, 193.
37. M. Meisel, *Z. Chem.*, 1983, **23**, 117.
38. D. Dubois, E. N. Duesler and R. T. Paine, *J. Chem. Soc., Chem. Commun.*, 1984, 488.
39. R. Blachnik and U. Wickel, *Angew. Chem., Int. Ed. Engl.*, 1983, **22**, 317.
40. M. Meisel and H. Grunze, *Z. Anorg. Allg. Chem.*, 1970, **373**, 265.
41. A. M. Griffin, P. C. Minshall and G. M. Sheldrick, *J. Chem. Soc., Chem. Commun.*, 1976, 809.
42. A. J. Burn, R. W. Joyner, P. Meeham and M. A. Parker, *J. Chem. Soc., Chem. Commun.*, 1986, 982.
43. H. W. Roesky, N. Benmohamed, M. Noltemeyer and G. M. Sheldrick, *Z. Naturforsch., Teil B*, 1986, **41**, 938.
44. R. A. Cherkasov, G. A. Kutyrev and A. N. Pudovik, *Tetrahedron*, 1985, **41**, 2567.
45. E. Thilo, *Angew. Chem., Int. Ed. Engl.*, 1965, **4**, 1039.
46. C. Lensch, W. Clegg and G. M. Sheldrick, *J. Chem. Soc., Dalton Trans.*, 1984, 723.
47. B. Cetinkaya, P. B. Hitchcock, M. F. Lappert, A. J. Thorne and H. Goldwhite, *J. Chem. Soc., Chem. Commun.*, 1982, 691.
48. C. Lensch and G. M. Sheldrick, *J. Chem. Soc., Dalton Trans.*, 1984, 2855.
49. E. Wiberg and E. Amberger, *Hydrides of the Elements of the Main Group IV*, Elsevier, Amsterdam, 1971.
50. S. Masumune, Y. Kabe, S. Collins, D. J. Williams and R. Jones, *J. Am. Chem. Soc.*, 1985, **107**, 5552.
51. J. S. Falcone, Jr (Ed.), *Soluble Silicates*, ACS Symposium Series, No. 194, American Chemical Society, Washington, DC, 1982.
52. F. Liebau, *Structural Chemistry of Silicates*, Springer Verlag, Berlin, 1985.
53. H. Berwe and A. Haas, *Chem. Ber.*, 1987, **120**, 1175.
54. A. Haas, H. Jurgen-Kutsch and C. Kruger, *Chem. Ber.*, 1987, **120**, 1045.

Note: See also *The Chemistry of Inorganic Homo- and Heterocycles*, Vols **I** and **II**, I. Haiduc and D. B. Sowerby, Academic Press (1987), London.

CHAPTER FOUR

Electron-rich Species

4.1 BORON–NITROGEN COMPOUNDS

Borazene, $B_3N_3H_6$, is the best known example of an inorganic 'aromatic' compound. It is isoelectronic with benzene and has often been refered to as 'inorganic benzene' because the physical properties of the parent compound and its methyl-substituted analogues are similar to those of benzene, toluene, etc. (Table 4.1).

4.1.1 Preparation

A number of preparative routes exist.[1,2]

1. Condensation of metal borohydrides with ammonium salts, or of boranes with amines, provides a useful route to N-substituted compounds:

$$3MBH_4 + 3RNH_3Cl \longrightarrow H_3B_3N_3R_3 + 3MCl + 9H_2$$
$$90\%$$

$$B_5H_9 + 3MeNH_2 \longrightarrow H_3B_3N_3Me_3 + B_2H_6 + 3H_2$$

$$RNH_2 + 1/2B_2H_6 \rightarrow RH_2NBH_3 \xrightarrow[-H_2]{100°C} H_6B_3N_3H_3R_3 \text{ (inorganic cyclohexane)}$$

$$80\% \text{ yield for } R = Me$$

$$\text{200°C/vacuum} \downarrow {}_{-H_2}$$

$$H_3B_3N_3R_3$$

$$B_4H_{10}.2NH_3 \xrightarrow{200°C} H_3B_3N_3H_3$$
$$40\%$$

Table 4.1 Boiling points of substituted borazenes and related aryl species ($N.B.$ borazenes are all N-substituted)

Borazene	B.p./°C	Aryl compound	B.p/°C
$B_3N_3H_6$	55	Benzene	80
$B_3N_3H_5Me$	84	Toluene	111
$B_3N_3H_4Me_2$	108	Xylene	130–140
$B_3N_3H_3Me_3$	133	Mesitylene	165

2. Condensation of boron halides with ammonium salts in high-boiling solvents such as chlorobenzene (N.B. this is similar to the route to phosphazene, see p. 80):

$$BX_3 + RNH_3X \xrightarrow[\text{cold}]{\text{fast}} RNH_3BX_4 \xrightarrow{\text{reflux}} RNH_2BX_3 \longrightarrow RNHBX_2$$

$$\downarrow$$

$$X_3B_3N_3R_3$$

$$X = Cl, R$$
$$= H; \ 50\% \text{ yield}$$

$$\downarrow \text{ LiAlH}_4$$

$$H_3B_3N_3H_3$$

3. Boron–oxygen compounds provide an alternative (and cheap) route:

$$RNH_2 + B(OPh)_3 + Al \xrightarrow[150°C]{H_2 \ 200 \text{ atm}} H_3B_3N_3R_3$$
$$70\%$$

4. The iminoborane $^iPrB \equiv N^tBu$ trimerizes to form $(^iPrBN^tBu)_3$, which has an unusual folded ring structure with two short $B{=}N$ double bonds (1.36 Å) and a cross-ring $B{-}N$ single bond (1.75 Å), and can thus be considered to be the equivalent of one of the possible canonical forms of benzene (Dewar benzene)[2a]. An interesting route to B_6N_7 (phenalene) was recently reported.[2] The synthesis relies (to a large extent) on the difference in $Sn{-}N$ and $Sn{-}S$ bond energies.

$$R = {}^iPr$$
$$R' = {}^tBu$$

5. Boron nitride, formally a ring compound, is prepared in high-temperature reactions, e.g. treatment of BCl_3 with NH_3 followed by pyrolysis at $750\,°C$.

6. Reactions to give related ring systems are also known. For example, substituted sulphur diimides and trithiadiboralanes[3] give mixed B—S—N rings, whilst mixed P—N—B rings can be prepared from $Me_2S.BH_3$ adducts.[4]

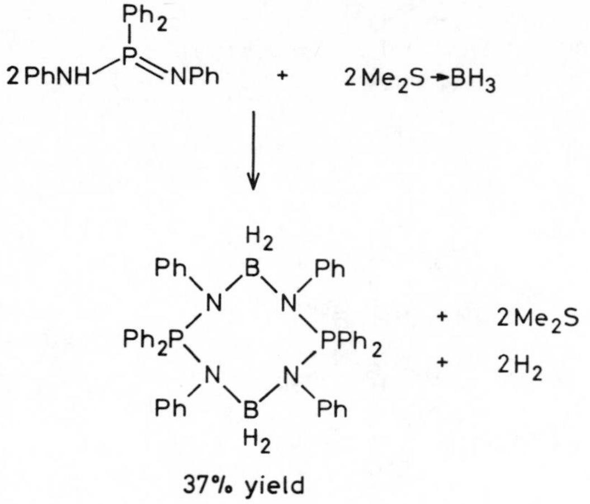

37% yield

4.1.2 Structure and Bonding

With the exception of (^{i}PrBN^{t}Bu)$_3$, mentioned above,[2a] borazenes are planar rings[5] with B—N distances of *ca* 1.43 Å, which compares with B—N single bond distances of 1.56 and 1.59 Å in $H_3N.BH_3$ and $H_6B_3N_3H_6$, respectively. An inexplicable feature is the B—H bond length in borazene itself (1.26 Å), which is long compared with those observed in boranes (1.1–1.2 Å). In the B_6N_7 (phenalene), mentioned above, the central B—N framework is planar but there is some distortion which is believed to be a result of the bulky $SiMe_3$ groups—the parent $B_6N_7H_9$ is not yet known and so the importance of π orbitals in larger rings is still open to question.

The B—N distances within the rings are clear evidence for a bond order greater than one and the UV photoelectron spectrum of borazene[6] is similar to that of benzene with the HOMO being a π orbital. The ^1H NMR spectrum consists of a triplet at $\delta = -5.4$ (protons on nitrogen split by $^{14}N, I = 1$) and a quartet at $\delta = 4.46$ (protons on boron split by $^{11}B, I = 3/2$).[7]

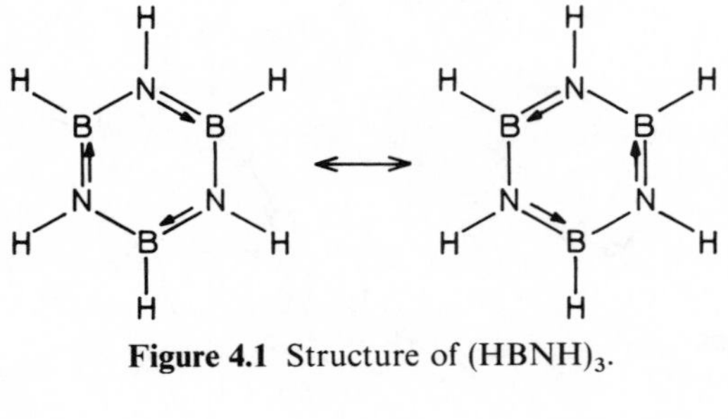

Figure 4.1 Structure of (HBNH)$_3$.

Formally, the 6π aromatic structure is represented by Fig. 4.1; however, this gives a misleading picture of the bond polarity. Although the nitrogen atoms are providing electrons for π orbitals, the σ orbitals are polar ($B \to N$), resulting in $B^{\delta+} - N^{\delta-}$. It is therefore to be expected that nucleophiles/Lewis bases will attack at boron whereas electrophiles/Lewis acids will attack at nitrogen.

Borazanes, isoelectronic with cyclohexane, have chair conformations.[8]

4.1.3 Reactions

The dominant feature in the reactivity (Fig. 4.2) of borazenes is bond polarity rather than aromaticity; for reviews, see refs 9 and 10. Solvolysis reactions with alcohols or amines result in ring cleavage or substitution. Halogen substitution of $H_3B_3N_3R_3$ occurs at boron via nucleophilic attack to give $X_3B_3N_3R_3$, the halogens of which can be readily substituted using alkyllithiums or Grignard reagents. Bulky R groups stabilize the ring for chloro compounds even against hydrolytic cleavage, but $H_3B_3N_3R_3$ are fairly sensitive to moisture. The formation of larger rings (naphthalene equivalents) and 6π metal complexes[11] such as $[M(CO)_3(borazene)]$ and $[Rh(COD)(borazene)]^+$ is a reminder of the available π electrons in these compounds.

4.2 PHOSPHORUS–NITROGEN COMPOUNDS

Phosphonitrillic chloride, $(PNCl_2)_3$, is the earliest reported inorganic heterocycle, dating back to 1834; it is the first in a large class of compounds.[12,13] Apart from being of commercial importance, these compounds pose interesting bonding problems. There are also various formally saturated systems, phosphazanes, and these are discussed first.

4.2.1 Phosphazanes

Illustrative examples together with preparative routes are shown in Fig. 4.3. Most of the reactions employed are straightforward, relying on the formation of ammonium halide salts or trimethylsilyl halides for their driving force. There are numerous examples of four-membered rings but only a few verified cases of six-membered rings.

Although the four-membered rings are formally saturated, there have been claims that they have some π character.[12] Support for $N \to P$ π donation comes from the observation that cyclic species are observed for compounds where the nitrogen has R groups which are electron-releasing and the phosphorus has electronegative substituents. Thus, for example, Ph_3PNH is a monomer whereas F_3PNH is a cyclic dimer. The observed $P-N$ bond lengths are in the range 1.69– 1.73 Å (close to single bonds). For a discussion of this problem and an excellent review of P^{III} phosphazanes, see ref. 14.

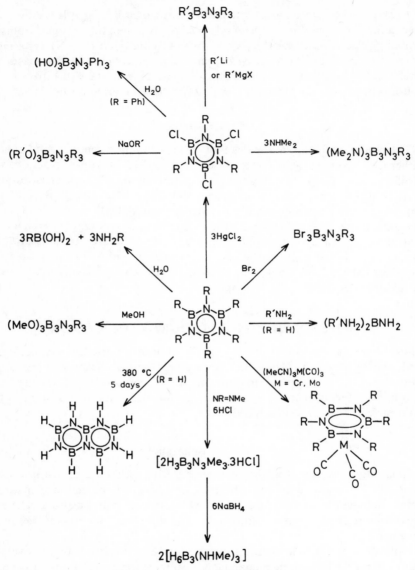

Figure 4.2 Reactions of borazenes.

4.2.2 Phosphazenes

4.2.2.1 Preparation

The range of phosphazene compounds known is very large (Fig. 4.4); various ring sizes and polymers are known.[12,15] Normally they are prepared by reaction

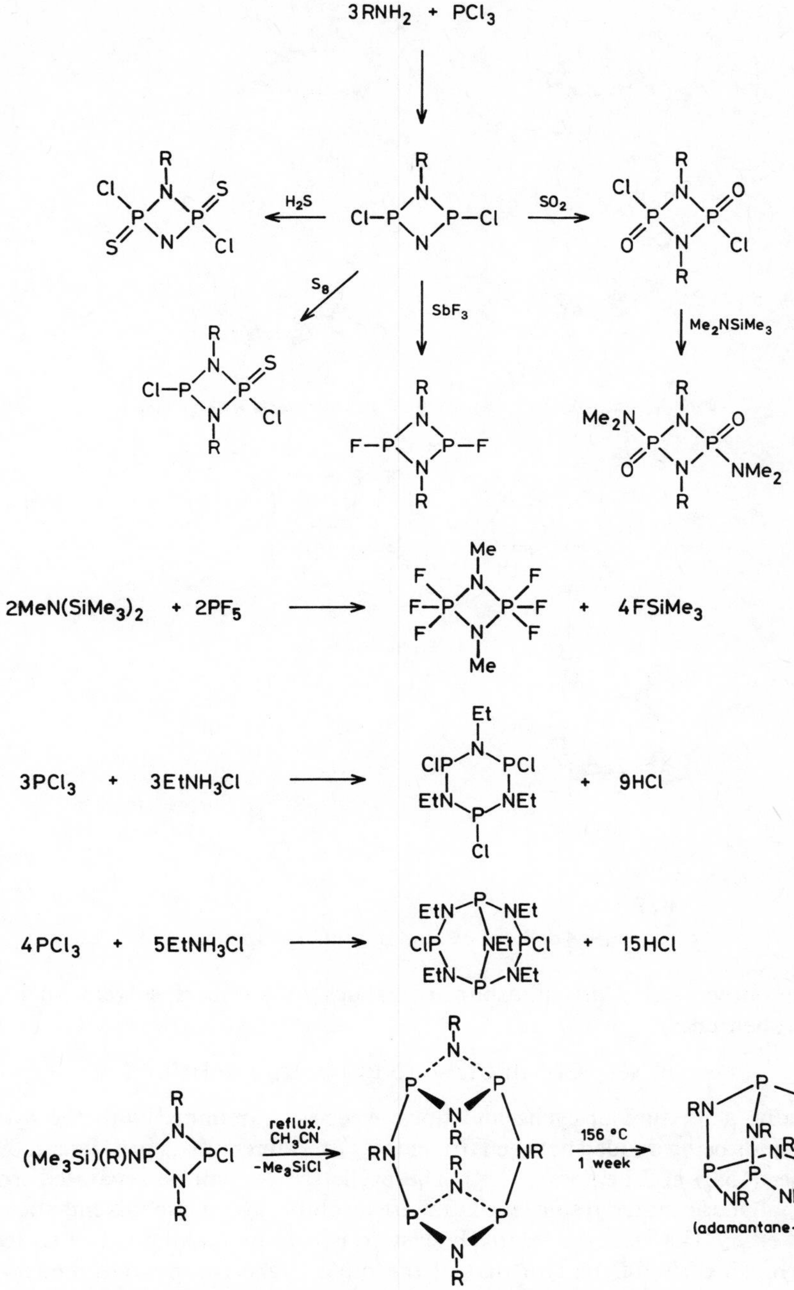

Figure 4.3 Preparative routes to phosphazanes.

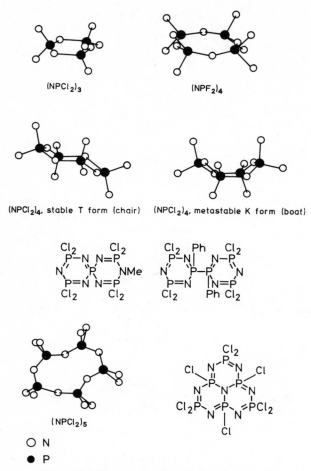

Figure 4.4 Some examples of phosphazenes.

of ammonium salts with phosphorus halides in an inert solvent such as dichlorobenzene:

$$n[NH_4]Cl + nPCl_5 \longrightarrow (NPCl_2)_{3-10} + 4nHCl$$

Generally, a mixture of cyclic and linear species is obtained with the cyclic compounds being divided between trimer (40%), tetramer (20%), pentamer (20%), hexamer (15%) and heptamer (5%). The cyclic species can be separated from linear polymeric material since the latter are insoluble in organic solvents such as diethyl ether. The trimer is relatively easy to isolate by making use of its base strength, which is different to that of the other cyclic compounds. Thus, the mixture is extracted into concentrated H_2SO_4, which is then diluted with water and back-extracted with light petroleum.

Bromo compounds are formed by using $PBr_3–Br_2$ mixtures rather than PCl_5. Alkyl/aryl-substituted species can be obtained from substitution reactions on $(NPCl_2)_3$ or directly by using $PRCl_4$ or PR_2Cl_3.

The mechanism of the reaction is not clearly understood. Some evidence for that shown below is the isolation of one of the proposed intermediates.

$$3PCl_5 + NH_4Cl \rightarrow [Cl_3P{=}N{=}PCl_3]PCl_6 + 4HCl$$

$$NH_4{}^+ + PCl_6{}^- \rightarrow Cl_3P{=}NH + 3HCl$$

$$Cl_3P{=}NH + Cl_3PNPCl_3{}^+ \rightarrow Cl_3P{-}PCl_2{=}N{=}PCl_3{}^+ + HCl$$
$$\downarrow {\small Cl_3P=NH}$$

$$\text{longer chain species} \leftarrow Cl_3P{=}(N{-}PCl_2)_2{=}N{=}PCl_3{}^+$$
$$\downarrow$$
$$(NPCl_2)_3 + PCl_4{}^+$$

An alternative route, making use of linear phosphoranes such as $H_2NP(Ph_2)NP(Ph_2)NH_2$ and its trimethylsilyl derivative, provides access to aryl- and mixed alkyl/aryl-substituted compounds (Fig. 4.5). A recent and elegant further application of this type of reaction is the formation of phosphazenes in which a phosphorus is replaced with a metal (W, Mo, Nb) or selenium/tellurium atoms.[16]

4.2.2.2 Structure and Bonding

The phosphazenes exist as both ring and chain compounds. Typically, their UV spectra are simply those of the substituents, i.e. there are no directly comparable transitions to those in benzene. The $v_{asym}(PNP)$ values in the IR spectra of the ring compounds are characteristic of various ring sizes. In attempting to discuss the bonding, there are a number of features to rationalize: (1) rings of different sizes, planar and puckered, are stable; (2) P—N bond lengths in rings are shorter than that expected for pure σ and there is no bond alternation; (3) nitrogens tend to be basic in character, especially when phosphorus is substituted by electron-rich species; (4) no organic π-type spectra are seen and, in contrast to organic aromatics, the PN skeleton is difficult to reduce electrochemically; (5) the bond angles are NPN 120° and PNP 120–150°.

The first, and most obvious, conclusion that we can reach is that the multiple bonding that exists is not directly analogous to that in organic aromatics. A number of bonding treatments have been proposed. In the zwitterionic model, resonance forms shown in Fig. 4.6 are used. The model is simple, without any need to utilize 3d orbitals, and is plausible on the basis of the electronegativity difference between N and P (N > P, 0.9); however, the chemical properties of phosphazenes are not consistent with a highly polar ring and having N^- requires the nitrogen atoms to adopt sp^3 hybridization with substantially smaller PNP bond angles than those observed. Alternatively, a weak 4p–2p π overlap has been

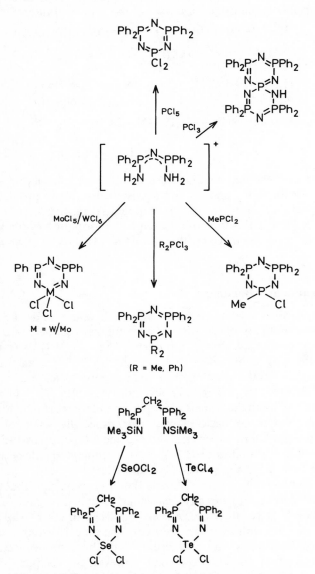

Figure 4.5 Synthetic routes to phosphazenes and metal complexes.

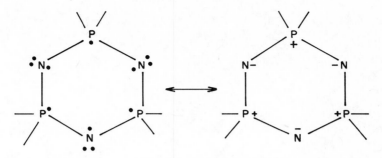

Figure 4.6 Zwitterionic bonding scheme in
$(NPCl_2)_3$.

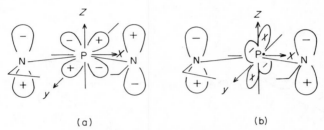

(a)   (b)

Figure 4.7 (a) Heteromorphic π_a and
(b) homomorphic π_a $d_\pi-p_\pi$ overlap.

suggested, but this type of system does not explain the stability of puckered rings.

The currently accepted treatment is based on $d_\pi-p_\pi$ overlap. Assuming sp^2-hybridized nitrogen and sp^3-hybridized phosphorus atoms, a simple σ framework in the ring and for P—Cl bonding is built up. This makes use of two nitrogen sp^2 orbitals and four phosphorus sp^3 orbitals; a further 'non-bonding' nitrogen sp^2 orbital is also filled. This leaves the p_z orbital on nitrogen and the d orbitals on phosphorus available for interaction. The p_z orbitals on nitrogen overlap with phosphorus d_{xz} orbitals to give an MO which has alternating phase round the ring (heteromorphic) or with d_{zy} to give a homomorphic MO. Since both of these are antisymmetric with respect to the plane of the ring they are labelled π_a (Fig. 4.7). If the homomorphic MO were the only important one, then we would expect the Hückel $4n + 2$ rule to apply whilst the heteromorphic system would result in a steady increase in π-bond energy with ring size and thus favour large rings. Since the ring stability is independent of ring size, a modified scheme using an orthogonal pair of orbitals derived from d_{xz} and d_{yz} (Fig. 4.8) is useful. This leads to π systems that are interupted at phosphorus and hence consist of 'islands.'[17]

Additionally, it is also useful to consider the sp^2 'lone pair' on nitrogen. It is

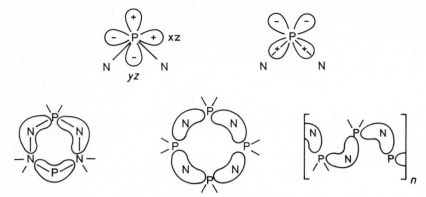

Figure 4.8 The orthogonal set of d orbitals and the resultant 'islands' of electron density.

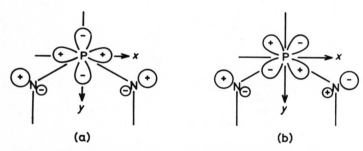

Figure 4.9 (a) Homomorphic π_s and (b) heteromorphic π_s overlap.

possible for this to donate electrons into the coplanar d orbitals on phosphorus (Fig. 4.9). Evidence for this type of interaction comes from the infrared spectra of $(PNX_2)_3$ rings; more electron-withdrawing groups on phosphorus result in higher frequencies for the $\nu_{asym}(PNP)$ vibration. In $(PNX_2)_3$ when X = Br, Cl or F, $\nu_{asym}(PNP)$ is 1175, 1218 and 1297 cm^{-1}, respectively.

It is clear that contributions of all the above types are necessary to explain fully the observed properties in phosphazene rings and chains. Indeed, it has also been suggested that cross-ring (transannular) P—P bonding may also have a significant contribution. For a more detailed discussion and some recent results of *ab initio* calculations, see ref. 18.

4.2.2.3 Reactions

MO calculations (CNDO/2) indicate, as expected on electronegativity grounds, that the phosphorus and nitrogen atoms in $(NPCl_2)_3$ are $\delta+$ and $\delta-$, respectively. The chemical properties (Fig. 4.10) are consistent with this. The major reactions are nucleophilic substitution of the halogens and Lewis base

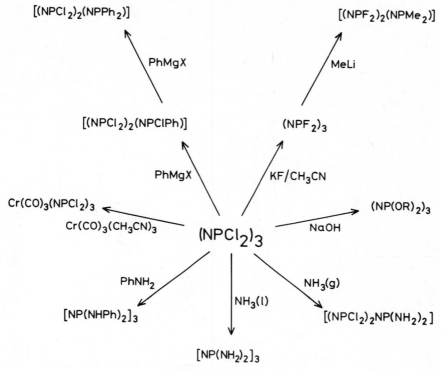

Figure 4.10 Reactions of $(NPCl_2)_3$.

behaviour at nitrogen. Replacement of chlorine by fluorine is accomplished by refluxing $(NPCl_2)_3$ with KF in acetonitrile; geminal substitution is favoured, i.e. 2, 2′, 4, 4′, 6, 6′. Substitution of chlorine by alkoxy, aryloxy or thiolate is straightforward, but $(NPCl_2)_3$ does not react with water (it can be steam distilled). Reaction with gaseous NH_3 gives $N_3P_3Cl_4(NH_2)_2$ whereas with liquid NH_3 complete substitution is effected. A wide range of organic amines have been reacted with $(NPCl_2)_3$ and the products obtained vary, in no readily interpreted way, depending on the amine. Even tertiary amines will substitute the chlorine atoms with elimination of RCl.

A wide range of organometallic reactions have been studied.[19] Alkyl-and aryllithiums substitute $(NPF_2)_3$, although excess MeLi causes ring degradation. The totally phenyl-substituted compound $(NPPh_2)_3$ is obtained in very low yield from reaction with PhMgBr; this reaction is very solvent dependent and the main product appears to be the monophenyl derivative. After prolonged Friedel–Crafts reaction using benzene–$AlCl_3$, under forcing conditions, the hexaphenyl compound is obtained. In view of the presence of a lone pair and a $\delta -$ charge on nitrogen, it is to be expected that these atoms will function as bases. The pK_a of

86

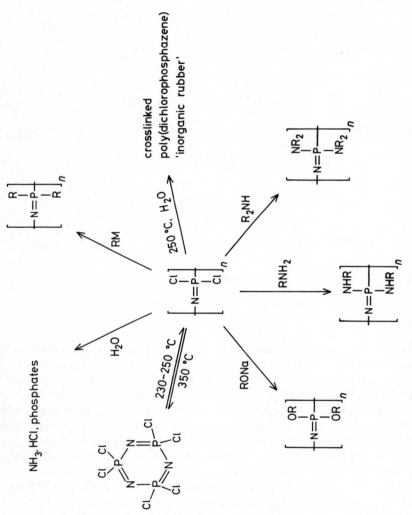

Figure 4.11 Polymerization reaction of $(NPCl_2)_3$.

substituted phosphazenes show a marked dependence on the nature of the substituent. The least basic compounds have electron-withdrawing groups (F or Cl, $pK_a = -6$) whereas electron-donating groups significantly enhance the basicity (NEt$_2$, $pK_a = 8$). Under the right conditions it is even possible to isolate protonated salts of phosphazenes.

Adducts with Lewis acids are known, particularly if the phosphorus substituent is electron-donating. For example, $(NPMe_2)_3$ forms a 1:1 adduct with SnCl$_4$. Metal complexes are also known; for example, $(NPMe_2)_3$ forms a simple monodentate complex with coordination via nitrogen when reacted with TiCl$_4$. Phosphazenes may also behave as bidentate ligands. Thus reaction of PtCl$_2$ with $[NPMe_2]_4$ gives $[NPMe_2]_4 \cdot PtCl_2$ in which the platinum is 2,6-nitrogen coordinated in a transannular fashion. Aminophosphazenes have the potential to coordinate through skeletal and substituent atoms. Thus, $[NP(NMe_2)_2]_4$ reacts

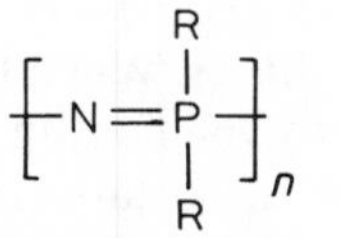

Linear poly(organophosphazene)s

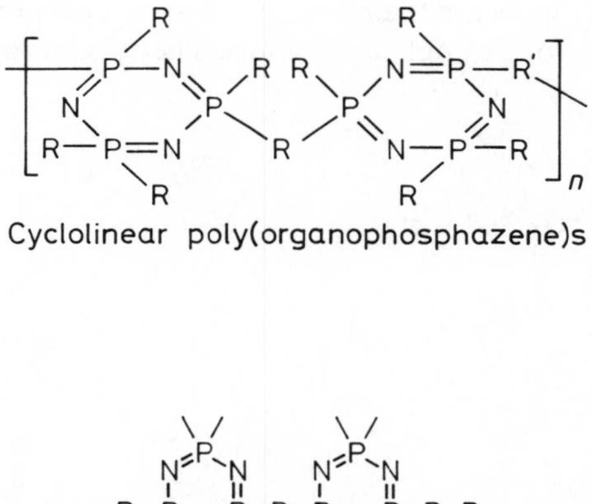

Cyclolinear poly(organophosphazene)s

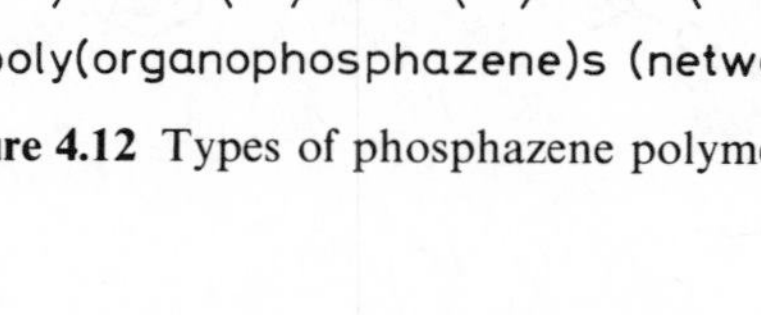

Cyclomatrix poly(organophosphazene)s (network polymers)

Figure 4.12 Types of phosphazene polymers.

with $Mo(CO)_6$ to give $[NP(NMe_2)_2]Mo(CO)_4$, in which the phosphazene acts as a bidentate ligand with the metal being coordinated by one ring and one side-group nitrogen. One might also expect π-complex formation and it has been reported that $(NPCl_2)_3$ reacts with $Cr(CO)_3(CH_3CN)_3$ to give a compound of formula $(NPCl_2)_3.Cr(CO)_3$. However, it is not known if the structure of this molecule is analogous to the organometallic analogue $Cr(CO)_3(C_6H_6)$.

Commercially the most important reaction of $(PNCl_2)_3$ is its polymerization. Heating to ca 230 °C under vacuum gives a linear polymer, $(NPCl_2)_n$. The formation of this polymer is reversible (Fig. 4.11) unless further heating in the presence of water or substitution of the halogens is carried out. Interestingly, attempts to polymerize $(NPR_2)_3$ have met with little success.

The three main types of polymer are illustrated in Fig. 4.12. Linear polymers $(NPR_2)_n$ ($n = 15$–$20\,000$) with a wide variety of substituents are known and these enable polymers suitable for many different applications to be prepared. For example, when $R = CF_3CH_2O$— or C_6H_5O— the polymer may be formed into a sheets or thin films or spun into fibres. When $R = NHR'$ water-soluble polymers are obtained; recently, a prototype for an anaesthetic high polymer ($R =$ procaine) was prepared. An important property of many phosphazene polymers is their fire resistance and stability over a wide temperature range, which makes them suitable for use in harsh conditions. Currently, they are used in O-rings, car components such as fuel lines and shock absorbers and as fire-resistant coatings. The usefulness of TCNQ salts of quaternized phosphazenes as conductors has recently been investigated.[20]

Cyclomatrix polymers are the result of excessive cross-linking between multifunctional systems, for example using diols. They are hard, high-melting resins.

4.3 SUBSTITUTED SULPHUR–NITROGEN RINGS

4.3.1 Preparation and Structure

Isoelectronic with $(NPX_2)_{3,4}$ are $(NSX)_3$ (X = F, Cl) and $(NSF)_4$. Oxidized S^{VI} compounds, $(SNOX)_3$, are also known and, as might be expected, mixed phosphazene–thiazene rings can be prepared (Fig. 4.13). Structurally, the

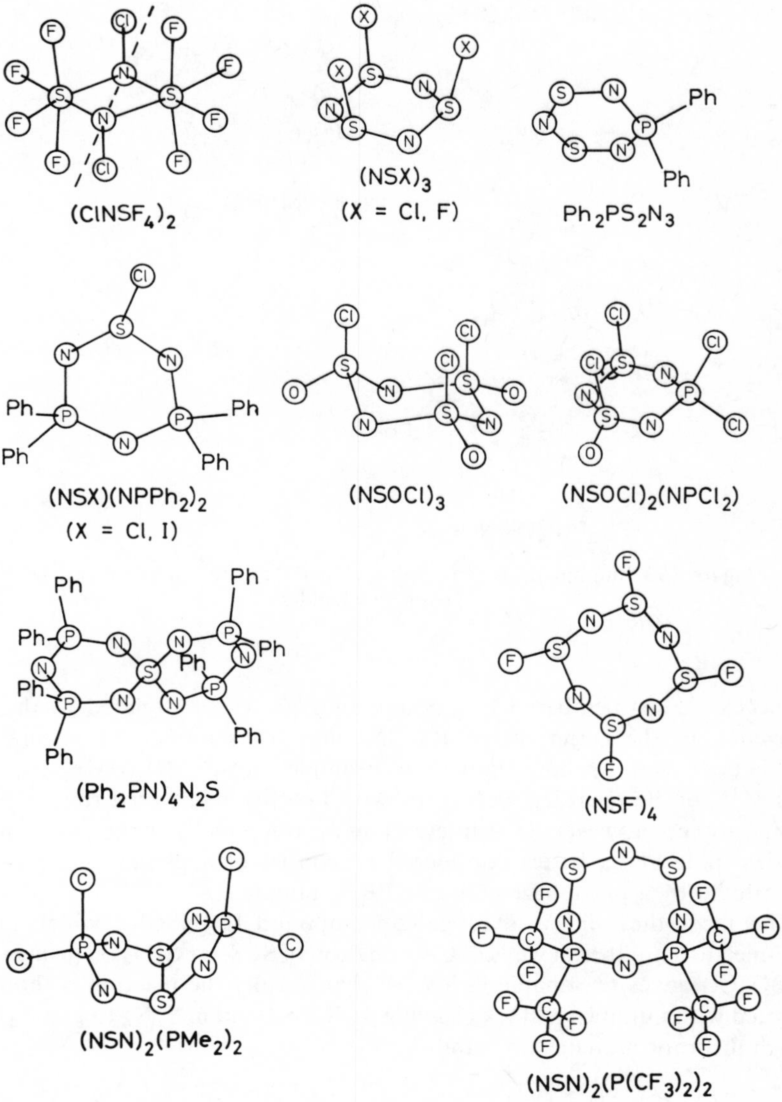

$(ClNSF_4)_2$

$(NSX)_3$
$(X = Cl, F)$

$Ph_2PS_2N_3$

$(NSX)(NPPh_2)_2$
$(X = Cl, I)$

$(NSOCl)_3$

$(NSOCl)_2(NPCl_2)$

$(Ph_2PN)_4N_2S$

$(NSF)_4$

$(NSN)_2(PMe_2)_2$

$(NSN)_2(P(CF_3)_2)_2$

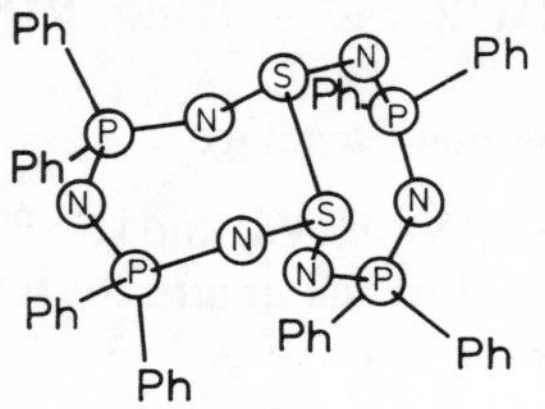

$(Ph_2PN)_4(SN)_2$

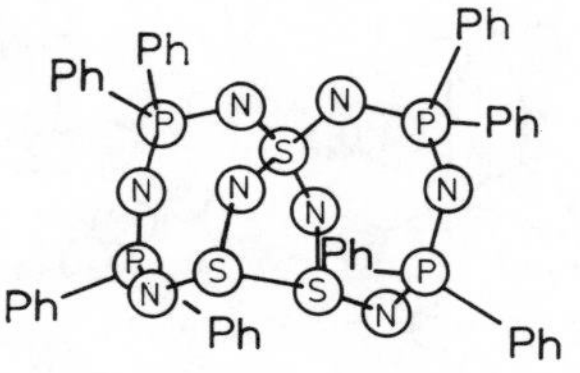

$(Ph_2PNPPh_2)_2S_3N_6$

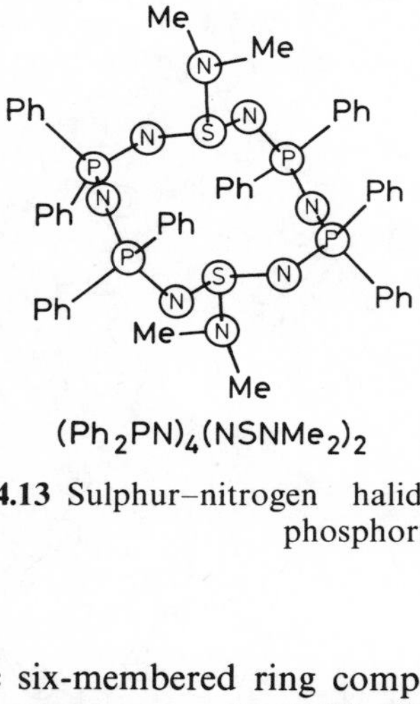

$(Ph_2PN)_4(NSNMe_2)_2$

Figure 4.13 Sulphur–nitrogen halides and mixed sulphur–nitrogen–phosphorus halides.

isoelectronic six-membered ring compounds are closely related to the phosphazenes. In the larger mixed PS–SN rings transannular (cross-ring) S—S interactions can be important. For example, 1, 5-$(NSN)_2(PMe_2)_2$ (S—S = 2.55 Å)[21] can be regarded as bicyclic with two five-membered rings sharing a common edge. However, S—S interactions are not possible in the 1, 3-isomer and so this molecule is better considered to consist of a planar N_3S_2 fragment substituted by a phosphazene-like $N(PR_2)_2$ unit.

The most thoroughly investigated compound is $(NSCl)_3$, which may be obtained in a number of ways.[22] Chlorination of S_4N_4 or $S_4N_4H_4$ (using Cl_2 gas or SO_2Cl_2) gives reasonable yields. Mechanistically the reaction is thought to proceed with initial Cl_2 attack cleaving an S—S bond in S_4N_4 to give $S_4N_4Cl_2$, which disproportionates on standing:

$$2S_4N_4 + 2Cl_2 \longrightarrow 2S_4N_4Cl_2 \longrightarrow 4/3(NSCl)_3 + S_4N_4$$

A more convenient synthesis[23] which does not involve explosive S_4N_4 is the chlorination of $[S_3N_2Cl]Cl$:

$$3[S_3N_2Cl]Cl + 3Cl_2 \longrightarrow 2(NSCl)_3 + 3SCl_2$$

Sulphanuric chloride, $(NSOCl)_3$, (named because of its relationship to cyanuric chloride) can be obtained in very low yield by oxidation of $(NSCl)_3$ with SO_3. A better route is the thermal decomposition of trichlorophosphazosulphuryl chloride (Cl_3PNSO_2Cl):[24]

$$2PCl_5 + H_3NSO_3 \longrightarrow 3HCl + POCl_3 + Cl_3P = NSO_2Cl$$
$$\downarrow \Delta$$
$$(NSOCl)_3 + 3POCl_3$$

This thermolysis has been reported to give rise also to traces of $(NSOCl)_2(NPCl_2)$.[25] Finally, reaction of thionyl chloride with sodium azide provides an alternative (although potentially dangerous) route:

$$SOCl_2 + NaN_3 \longrightarrow (N_3SOCl) + NaCl \longrightarrow 1/3(NSOCl)_3 + N_2$$

Treatment of S_4N_4 with dilute F_2 or AgF_2 gives $(NSF)_4$ in ca 12% yield. Fluorination of $(NSCl)_3$ with AgF_2 gives $(NSF)_3$ in 90% yield. Sulphanuric fluoride is obtained by treating $(NSOCl)_3$ with KF in an analagous fashion to the conversion of $(NPCl_2)_3$ to $(NPF_2)_3$.

There is one example[26] of a four-membered ring, $(ClNSF_4)_2$:

$$3NSF_3 + 4ClF \longrightarrow F_3SNCl_2 + (ClNSF_4)_2 + F_2$$
$$3.6\%$$

Mixed PN–SN rings can be prepared in a number of ways[27] and illustrative examples are shown below. One feature which is immediately apparent with the mixed P—S—N rings is the increased diversity, larger rings and spiro compounds being known.

4.3.2 Reactions

Trithiazyl trichloride, $(NSCl)_3$, is a moisture-sensitive, yellow solid which dissociates with some decomposition to NSCl when heated. In donor solvents such as thf it behaves much like the monomer NSCl.thf, although the exact nature of the solutions is not clear. Dehalogenation is readily accomplished. For example, treatment with mercury gives S_4N_4. Reaction with silver salts gives an *in situ* source of the NS^+ cation and has been used in the formation of thionitrosyl complexes.[28] The NS^+ cation reacts with organic solvents and therefore a related reaction using $AgAsF_6$ in liquid SO_2 which gives the product as an isolable solid in 75% yield has been developed:[29]

$$(NSCl)_3 + 3AgAsF_6 \xrightarrow{CH_3CN} [NS][AsF_6] + 3AgCl$$

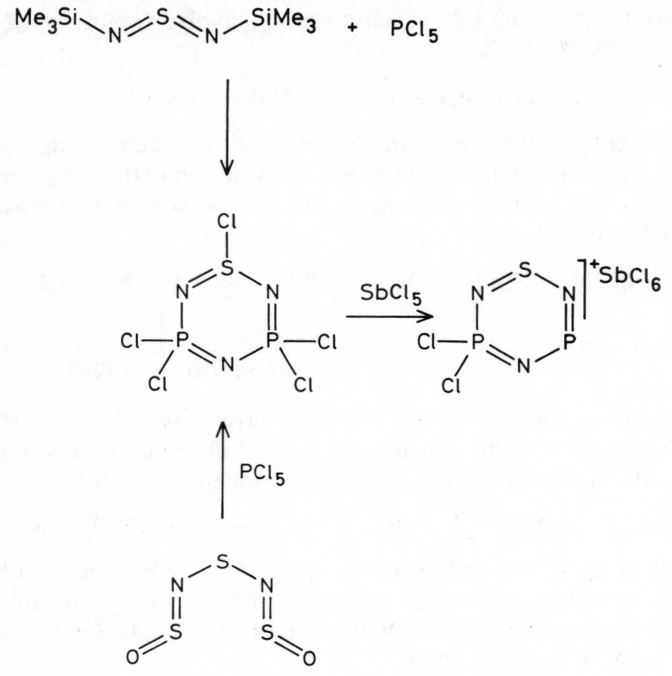
+ PCl5
SbCl5
PCl5

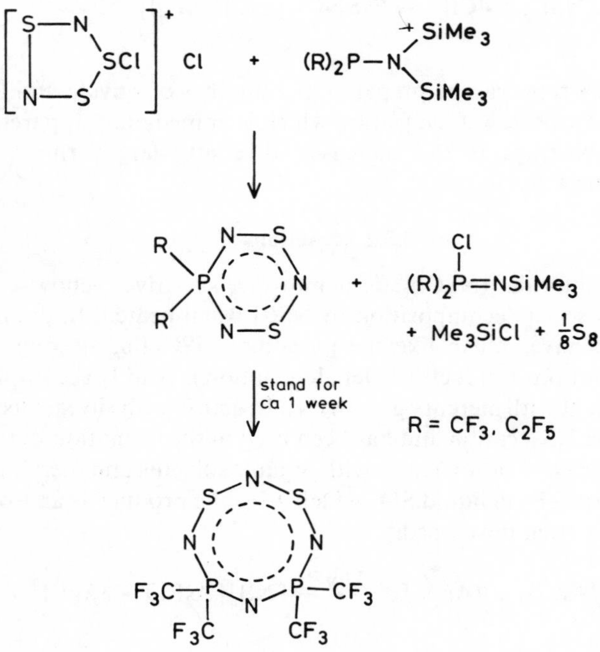
stand for
ca 1 week
R = CF3, C2F5

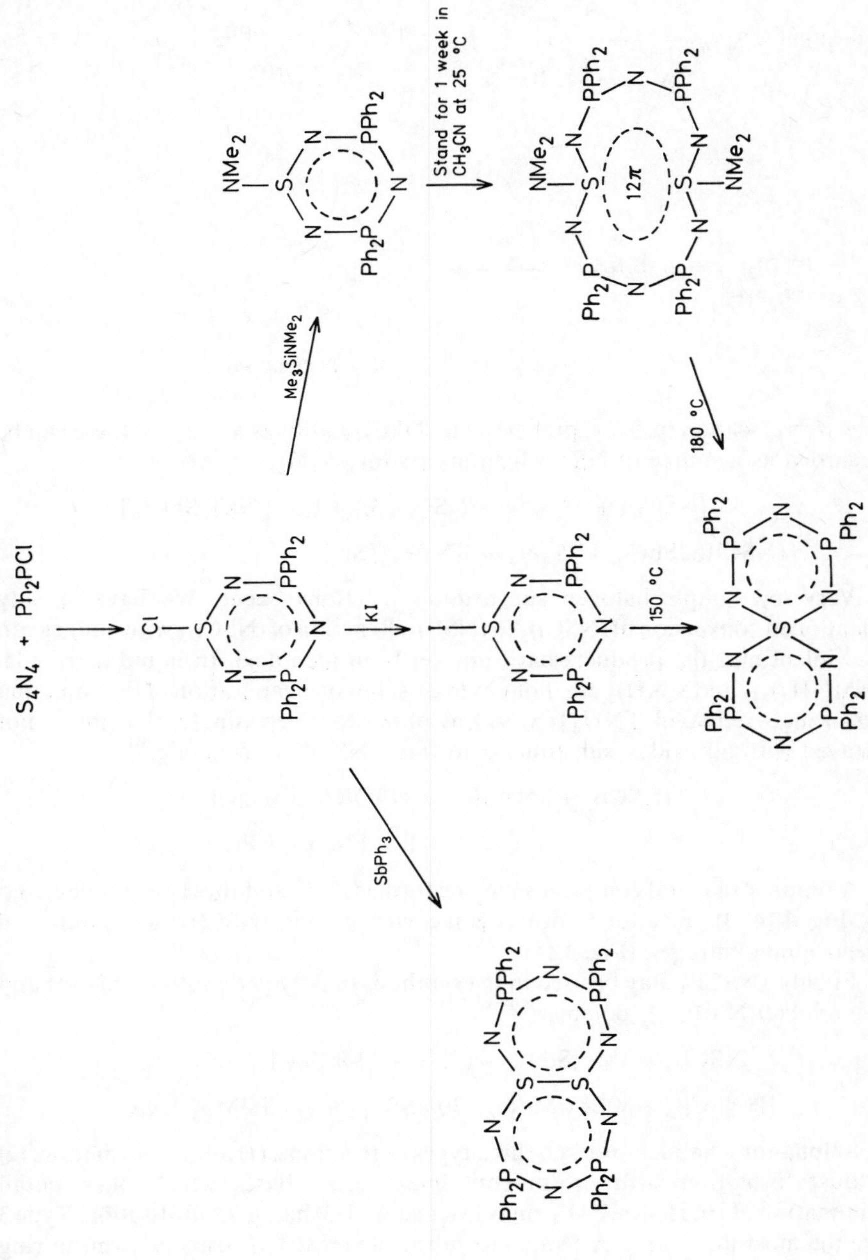
S4N4 + Ph2PCl
Me3SiNMe2
Stand for 1 week in CH3CN at 25 °C
12π
180 °C
KI
150 °C
SbPh3

$$Ph_2PPh_2 + S_4N_4 \longrightarrow$$

$$(PhO)_3P \quad + \quad S_4N_4 \longrightarrow$$
$$or \quad Ph_2PH$$

R = PhO or Ph

The $S_5N_5^+$ cation (p. 97) is prepared via $(NSCl)_3$ adducts and again these can be regarded as a source of NS^+ which inserts into S_4N_4:

$$(NSCl)_3 + 3SbCl_5 \rightarrow (NSCl)_3.3SbCl_5 \equiv [NS][SbCl_6]$$

$$(NSCl)_3.3SbCl_5 + 3S_4N_4 \rightarrow 3[S_5N_5][SbCl_6]$$

Very few simple halogen substitution reactions occur. We have already mentioned conversion of $(NSCl)_3$ to $(NSF)_3$. Reaction of $(NSCl)_3$ with amines can be violent and the products have not yet been identified. In liquid ammonia, $(NS(NH_2)_2)_3$ and $S(NH)_2$ are thought to exist but on evaporation of the ammonia an orange–red solid, $[NH_4][S_4N_5]$, is obtained. Surprisingly, the ring is not cleaved with alkoxides, substitution to form $(NSOR)_3$ occurring:[30]

$$(NSCl)_3 + 3NaOR \longrightarrow (NSOR)_3 + 3NaCl$$

$$R = Me, Et, i\text{-}Pr, etc.$$

A number of metal complexes may be formed,[31,32] and these are summarized in Fig. 4.14. In reactions with organic compounds $(NSCl)_3$ is a source of electrophilic nitrogen (Fig. 4.15).

Finally, $(NSCl)_3$ may be used in the synthesis of polymeric sulphur nitride and the related $(NSBr_{0.4})_x$ polymer:[33]

$$(NSCl)_3 + 3Me_3SiN_3 \longrightarrow (SN)_x + 3Me_3SiCl + 3N_2$$

$$10(NSCl)_3 + 30BrSiMe_3 \longrightarrow 30NSBr_{0.4} + 30ClSiMe_3 + 9Br_2$$

Sulphanuric halides undergo three types of reactions: (1) adduct formation; (2) adduct formation with proton-containing Lewis bases which often entail elimination of HCl followed by ring cleavage; and (3) halogen substitution. Type 3 are the most important. Attempts to substitute $(NSOCl)_3$ usually result in ring

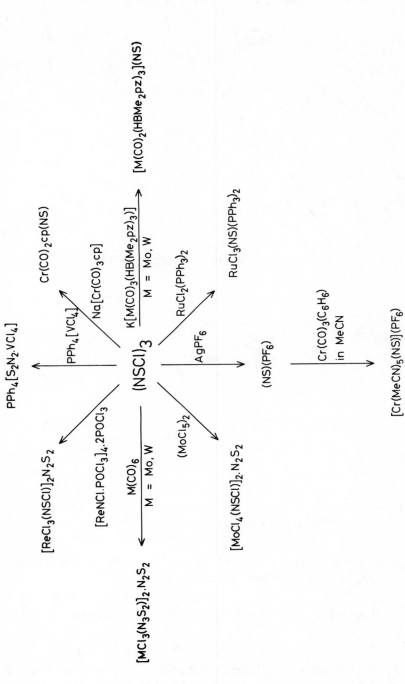

Figure 4.14 Reactions of $(NSCl)_3$ with metal halides.

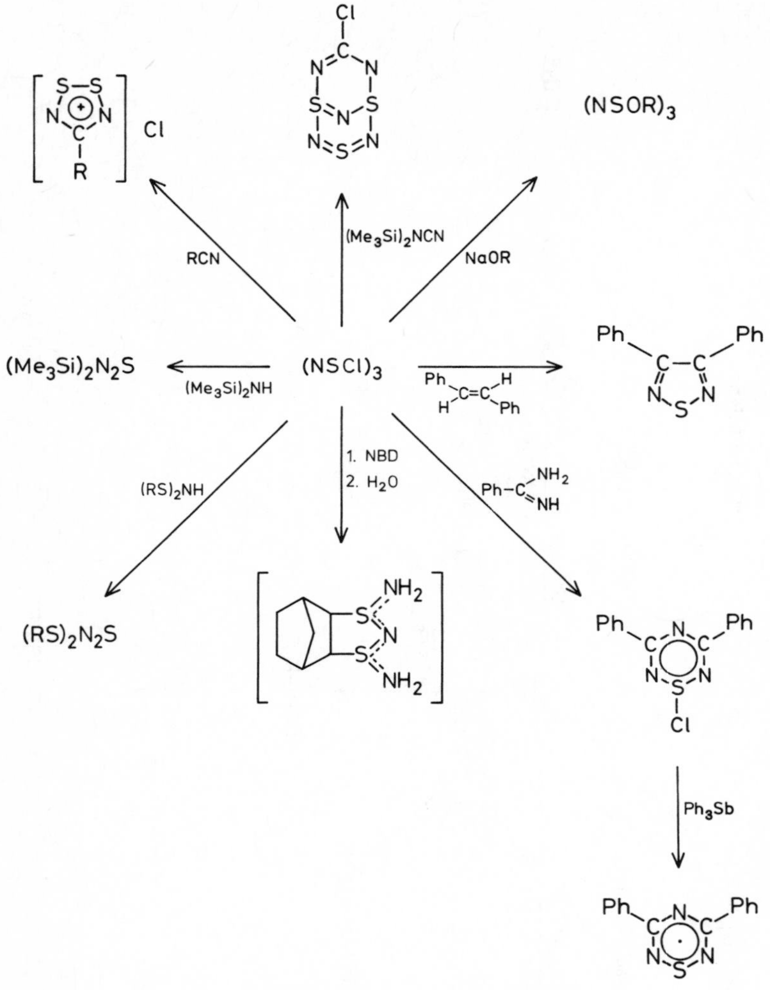

Figure 4.15 Organic reactions of (NSCl)$_3$.

cleavage and so most substitutions are carried out on the fluoride. The most important reactions are summarized in Fig. 4.16.[34]

4.4 PLANAR SULPHUR–NITROGEN SPECIES

4.4.1 Introduction

Using a simple extension of Hückel rules, Banister[35] has proposed that many planar sulphur–nitrogen species can be regarded as *formally* aromatic. It is

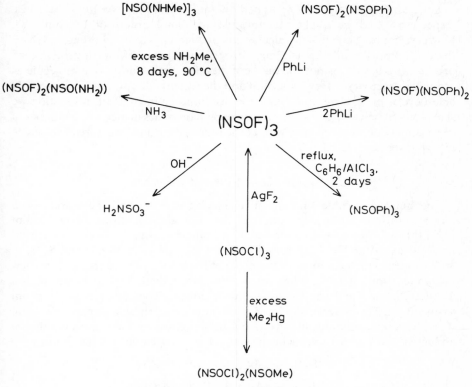

Figure 4.16 Reactions of sulphanuric halides.

Table 4.2 Pseudo-aromatic planar sulphur–nitrogen rings (those underlined are known to exist)

	Anions	Neutral	Cations	Others
6π	NS_2^-	S_2N_2	$S_3N_2^{2+}$	$S_3N_2Cl^+, S_3N_2O$
				5-membered C—S—N rings
7π			$S_3N_2^+$	**5-membered C—S—N rings**
10π	$S_3N_3^-$	S_3N_4, S_4N_2	$S_3N_5^+, S_4N_4^{2+}$	$S_3N_3O, S_4N_4O_2$
			$S_4N_3^+, S_5N_2^{2+}$	**7-membered C—S—N rings**
				8-membered C—S—N rings
14π		S_4N_6, S_5N_4	$S_5N_5^+, S_5N_6^{2+}, S_6N_3^+$	
18π	$S_6N_5^-$	S_6N_6	$S_6N_7^+, S_7N_6^+$	

assumed that each sulphur provides $2e^-$ and each nitrogen $1e^-$ into the π orbitals. Thus, S_2N_2 would be a 6π system whilst $S_4N_3^+$ and $S_3N_3^-$ would be a 10π systems. Table 4.2 classifies planar S–N species on the basis of the number of π electrons they contain. This simplistic approach for counting electrons leads to the conclusion that the bond orders in SN rings should be higher than in their organic analogues. For example, benzene and $S_3N_3^-$ are 6π and 10π rings, respectively, and we might therefore assume bond orders of 1.5 and 1.83. However, more detailed MO calculations predict a lower bond order in $S_3N_3^-$ than in benzene with the sulphur–nitrogen ring having four electrons in π^* orbitals. Bond order is not the only criterion for classification of a molecule as aromatic; reactivity is very important and the recently discovered organo-S—N heterocycles provide an opportunity to test bonding ideas.[36] In this section we shall limit ourselves to a brief discussion of planar species in increasing number of π electrons. For an excellent review of the preparation and structure of sulphur–nitrogen species, see ref. 37.

4.4.2 Four-Membered Rings

The smallest isolable sulphur–nitrogen ring is S_2N_2 and there are now a number of routes for its preparation.[22] The original method was sublimation of S_4N_4 through silver wool under a high vacuum (Fig. 4.17). However, improved methods are available. For example, the reaction of $S_2N_2.2AlCl_3$ and S_4N_4 at 80 °C under high vacuum gives the product in *ca* 40% yield. There are very few known reactions of S_2N_2. When recrystallized from diethyl ether the compound is unpredictable and may detonate with considerable force even at room temperature. The material sublimed from the vapour phase is supposedly more stable but understandably little work has been performed. There are a number of adducts known with Lewis acids in which the nitrogen acts as the donor:

$$S_2N_2 + 2SbCl_5 \longrightarrow S_2N_2.2SbCl_5$$

$$S_2N_2 + 4AlCl_3 \longrightarrow S_2N_2.2AlCl_3$$

$$2(NSCl)_3 + TiCl_4 \longrightarrow S_2N_2.TiCl_4$$

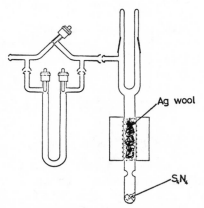

Figure 4.17 Apparatus for the preparation of S_2N_2 by pyrolysis of S_4N_4.

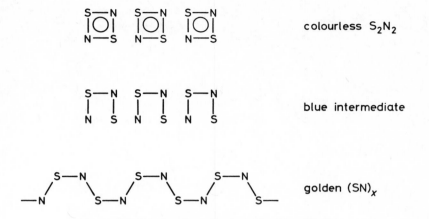

Figure 4.18 Mechanism of polymerization of S_2N_2.

A reaction which has caused considerable excitement in recent years is the solid-state polymerization of S_2N_2 to $(SN)_x$. On allowing S_2N_2 to warm to *ca* $-10\,^\circ$C, the colourless solid first becomes dark blue and then gradually changes to golden crystals of $(SN)_x$ during the course of 2–3 days. The mechanism of this polymerization has been studied by ESR; neither S_2N_2 nor $(SN)_x$ has an ESR signal but the dark intermediate gives a free radical signal ($g = 2.005$). It is assumed that a free radical S_2N_2 species is initially formed by bond rupture and that this species goes on to attack another S_2N_2 molecule in the crystal. This polymerization process is illustrated schematically in Fig. 4.18; it is noticeable that only small structural changes, i.e. widening in bond angles, are occurring although the resulting increase in stability is remarkable.

Polymeric sulphur nitride, $(SN)_x$, has a number of unusual features which make it of special interest. It exists as fibrous strands and conducts electricity along the polymer chain direction, but is an insulator perpendicular to the chains. Apart from being an anisotropic conductor, $(SN)_x$ was the first non-metal compound in which superconductivity (at 0.33 K) was established. The one-dimensional conductivity can be readily understood even in a simple valence-bond description of the bonding in $(SN)_x$—it is clear that more than one canonical form is possible. The preparation and properties of $(SN)_x$ have been reviewed.[38]

4.4.3 Five-Membered Rings

Synthetically, one of the easiest compounds to prepare in S—N chemistry is undoubtedly $[S_3N_2Cl]Cl$. In a solventless reaction:

$$4S_2Cl_2 + 2NH_4Cl \longrightarrow [S_3N_2Cl]Cl + 8HCl + 5S$$

ammonium chloride and disulphur dichloride are combined in a round-bottomed flask equipped with an air condenser. Usually sulphur is added to

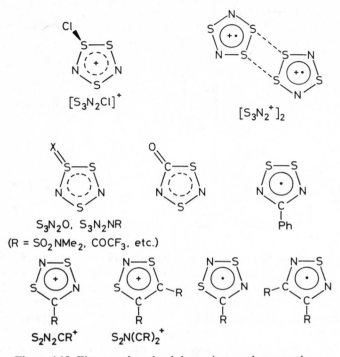

Figure 4.19 Five-membered sulphur–nitrogen heterocycles.

reduce the evolution of chlorine from the S_2Cl_2. Heating the reaction to a gentle reflux results in the formation of orange crystals of $[S_3N_2Cl]Cl$ in the air condenser in up to 60% yield. Its structure consists of a nearly planar ring with the chlorine above the plane. The S—N distances in the ring vary between 1.54 and 1.62 Å, indicating substantial π-bond character.

$[S_3N_2Cl]Cl$ is only poorly soluble in organic solvents and it reacts with polar solvents (SO_2Cl_2, formic acid) in which it dissolves. Nonetheless, it is a useful starting material in the synthesis of a number of other heterocycles (Fig. 4.19) and also $(SN)_x$. For example, the material (prepared as described above) can be used directly to make $(NSCl)_3$ or $[S_4N_3]Cl$. Other five-membered rings may also be prepared. Slow addition of a dichloromethane solution of anhydrous formic acid to $[S_3N_2Cl]Cl$ gives S_3N_2O with elimination of HCl and carbon monoxide. These two rings are isoelectronic ($Cl^- \equiv O^{2-}$). Reaction of $[S_3N_2Cl]Cl$ with sulphamide gives $S_4N_2O_2$:

$$4[S_3N_2Cl]Cl + 2SO_2(NH_2)_2 \longrightarrow 2S_4N_4O_2 + [S_4N_3]Cl + NH_4Cl$$
$$+ S_2Cl_2 + 4HCl$$

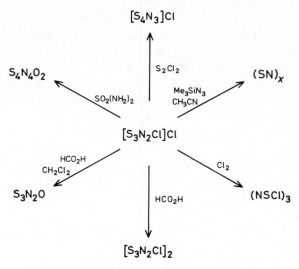

Figure 4.20 Reactions of [S₃N₂Cl]Cl.

If $[S_3N_2Cl]Cl$ is stirred with anhydrous formic acid (in the absence of solvent), a compound of empirical formula S_3N_2Cl is obtained. This material exists in the solid state as the dimer of the monocation $[S_3N_2]^+$ (Fig. 4.20) and there are numerous other examples with different counter ions. In solutions the monomeric radical cation has been observed by ESR. For example, treatment of S_4N_4 or $[S_3N_2Cl]Cl$ with $AlCl_3$ in CH_2Cl_2 rapidly (*ca* 1 s) results in a five-line spectrum with $g = 2.01$, indicating the presence of two equivalent nitrogen atoms ($I = 1$). In one case the monomeric cation has been isolated in the solid state. Reaction of S_4N_4 with $[Te_6^{4+}][AsF_6]_4$ gives $[S_3N_2][AsF_6]$.

There are several five-membered organo-S—N heterocycles. The most interesting of these are probably the thermally stable 7π paramagnetic liquids prepared as shown below.[39]

4.4.4 Six-Membered Rings

These species (all formally containing ten π electrons) are shown in Fig. 4.21. The most symmetric example is the $S_3N_3^-$ anion, which has close to trigonal bond angles at sulphur (117°) and nitrogen (123°) and S—N bond distances within the range 1.58–1.63 Å. A π MO scheme for the bonding in $S_3N_3^-$ is shown in Fig. 4.22. The most significant difference between this and the MO for benzene

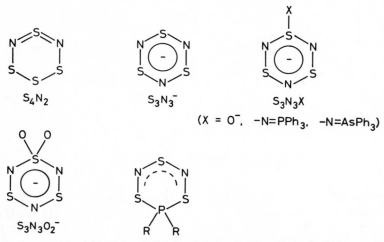

Figure 4.21 Six-membered sulphur–nitrogen heterocycles.

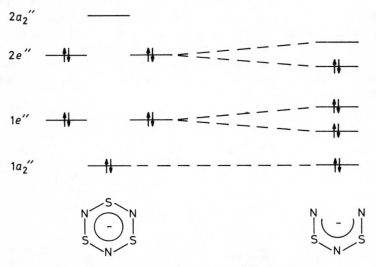

Figure 4.22 MO scheme for the bonding in the $S_3N_3^-$ anion.

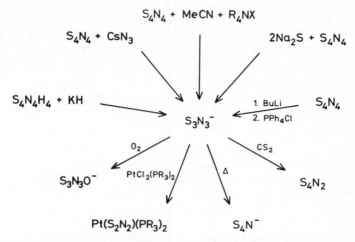

Figure 4.23 Preparative routes and some reactions of the $S_3N_3^-$ anion.

is that the two degenerate π^*-2e'', orbitals are filled in the case of the $S_3N_3^-$ anion, leading to a lower bond order and weaker framework bonding in the case of the SN species. The electronic spectrum of $S_3N_3^-$ consists of a strong absorption at *ca* 360 nm (ε is dependent on the counter ion and solvent) which is assigned to a $\pi^* \to \pi^*$ transition.[40]

The $S_3N_3^-$ ion is observed from a variety of reactions (Fig. 4.23).[41] It has been suggested that the formation of $S_3N_3^-$ from S_4N_4 proceeds via nucleophilic attack at sulphur, leading to ring opening to give a poly(sulphur–nitrogen) chain which cyclizes to give the final product. The preparation by deprotonation of $S_4N_4H_4$ with base may involve short-chain anions such as $S_2N_2^{2-}$ (currently only known in metal complexes). Although a great deal of work has been carried out in this area in recent years, it seems likely that there is a great deal of progress still to be made in understanding the solution equilibria that are occurring. For example, the S_4N^- anion is obtained by thermal decomposition of $S_3N_3^-$, by deprotonation of S_7NH and is also observed in solutions of sulphur in liquid ammonia.[42]

Historically better established than $S_3N_3^-$ is S_4N_2 (related by replacement of N^- by S). Tetrasulphur dinitride[43] consists of a planar S_3N_2 fragment with the remaining sulphur atom being above this plane (the so called 'half-chair'). The bond distances in S_4N_2 vary considerably around the ring. The most appropriate description is a sulphur diimide fragment (N_2S) bonded to an S_3 chain.

S_4N_2 is obtained from a number of reactions. Pyrolysis of mixtures of S_8 and S_4N_4 in a sealed tube is probably the most dangerous. More useful on a laboratory scale is the reaction of S_2Cl_2 in CS_2 with aqueous ammonia.

Alternatively, thermal decomposition of $Hg(S_7N)_2$ at room temperature gives the compound in 64% yield based on nitrogen. It exists as an evil-smelling red oil at room temperature and decomposes explosively at *ca* 100 °C and more slowly at room temperature. Currently, there are few reported reactions apart from addition to norbornadiene shown below.

Finally, in this section, are mixed PS–NS rings. Formally S^- is isoelectronic with PR_2 and so the existence of $S_3N_2(PR_2)$ rings[41] is not surprising. Figure 4.22 shows the MO scheme for $S_3N_3^-$ and the hypothetical $S_2N_3^-$ fragment which suggests that even with a heteroatom, six-membered rings will still have appreciable π stabilization.

Substituted six-membered rings are known. Oxidized forms of $S_3N_3^-$ are prepared by bubbling dry oxygen through solutions of $S_3N_3^-$. Reaction of triphenylphosphine/arsine with S_4N_4 proceeds with ring contraction to give a six-membered ring[44] which is itself thermally unstable (see below) and gives a substituted $S_3N_2^-$ chain[45] on heating (reminiscent of the product obtained on heating $S_3N_3^-$).

4.4.5 Seven-Membered Rings

The $S_4N_3^+$ cation has been known since the turn of the century. It is believed to be an intermediate in the formation of S_4N_4 and (like S_4N_4) is often obtained as an unwanted side-product. The preferred method is the reaction of easily obtainable $[S_3N_2Cl]Cl$ by refluxing with S_2Cl_2 in CCl_4:

$$3[S_3N_2Cl]Cl + S_2Cl_2 \longrightarrow 2[S_4N_3]Cl + 3SCl_2$$

The cation is planar with short S—N bonds (1.55 Å) and the UV–visible spectrum is consistent with a 10π delocalized system. The poor solubility of salts

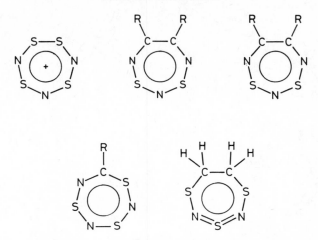

Figure 4.24 Seven-membered sulphur–nitrogen heterocycles.

of $S_4N_3^+$ means that there are few reported reactions, but seven-membered
organic substituted heterocycles have recently been discovered.[36]

Figure 4.24 summarizes the known seven-membered compounds. Originally,
the organo-S—N compounds were obtained in low yields from reactions of
acetylenes with S_4N_4, but more elegant routes, (e.g. as shown below) have been
developed.

DDQ = dichlorodicyanobenzoquinone

The carbons in trithiadiazapine undergo normal electrophilic substitution
reactions. They may be brominated with N-bromosuccinimide, and with
copper(II) nitrate trihydrate the mononitro compound is formed. Unlike many
S—N compounds there is little tendency to undergo cycloaddition reactions
with alkenes or dienes. Disappointingly, Friedel–Crafts reactions catalysed by
$AlCl_3$ do not occur, possibly because of coordination of the $AlCl_3$.

106

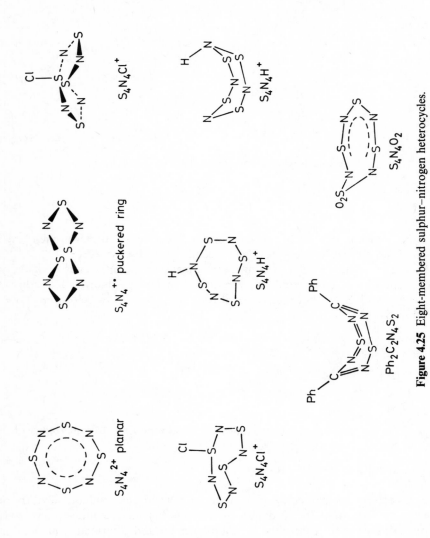

Figure 4.25 Eight-membered sulphur–nitrogen heterocycles.

4.4.6 Eight-Membered and Larger Rings

Figure 4.25 illustrates the more important examples of eight-membered rings. The 10π $S_4N_4{}^{2+}$ cation obeys the Hückel rule and is planar (D_{4h} symmetry), whilst increasing the number of π electrons results in the puckered ring of the $S_4N_4{}^{+\cdot}$ (11π) radical cation. The 12π $S_4N_4H^+$ has an even less planar structure which is similar to that observed for Lewis acid adducts of S_4N_4 (p. 113). Both $S_4N_4{}^{2+}$ and $S_4N_4{}^{+\cdot}$ have almost equal S—N bond lengths within the ring (1.55 and 1.54 Å, respectively), but bond alternation exists in $S_4N_4H^+$ (distances vary between 1.539 and 1.656 Å). These three compounds (together with S_4N_4) illustrate the structural variations possible in the same ring system as a result of changing the electron count.

Routes to eight-membered rings are shown below:

$$4(NSCl)_3 + 3FeCl_3 \xrightarrow{CCl_4} [S_4N_4Cl]FeCl_4 + 2[S_4N_4{}^\cdot]^{+\cdot}[FeCl_4] + 4Cl_2$$

$$75\%$$

$$S_4N_4 + \text{excess } SbCl_5 \xrightarrow{\text{liq. }SO_2} S_4N_4.SbCl_5 \rightarrow [S_4N_4][SbCl_6]_2$$

$$+ SbCl_3$$

Oxidation of S_4N_4 to give the dication is carried out under similar conditions to those used in the preparation of sulphur and selenium cations (p. 115). We have already mentioned the preparation of $S_4N_4O_2$ from $[S_3N_2Cl]Cl$ and sulphamide (p. 100).

Eight-membered C—S—N rings are also known, e.g. from the reaction of amidines with sulphur dichloride[46] in the presence of the base dbu (1,8-Diazabicyclo[5,4,0]undec-7-ene). The crystal structure with R = Et or Ph reveals planar rings, in contrast to examples in which R = NMe_2, which are folded. For a discussion of these and related C—S—N rings, see Gmelin.[36]

$$2RC(NH)NH_2 + 3SCl_2 \longrightarrow (RC)(NSN)_2(CR)$$

The largest currently known (planar) SN ring is the 14π $S_5N_5{}^+$ cation.[47] A wide variety of salts are known and the structure of the cation appears to depend on the anion. Perhaps the most easily understood synthetic strategy is the reaction of NS^+ with S_4N_4. The NS^+ cation is often prepared *in situ* from $(NSCl)_3$:

$$S_4N_4 + [NS][SbF_6] \xrightarrow{SOCl_2} [S_5N_5][SbF_6]$$

$$1/3(NSCl)_3 + AlCl_3 + S_4N_4 \longrightarrow [S_5N_5][AlCl_4]$$

$$(Me_3Si)N{=}S{=}N(SiMe_3) + FSO_2N{=}S{=}O \longrightarrow [S_5N_5][S_3N_3O_4]$$

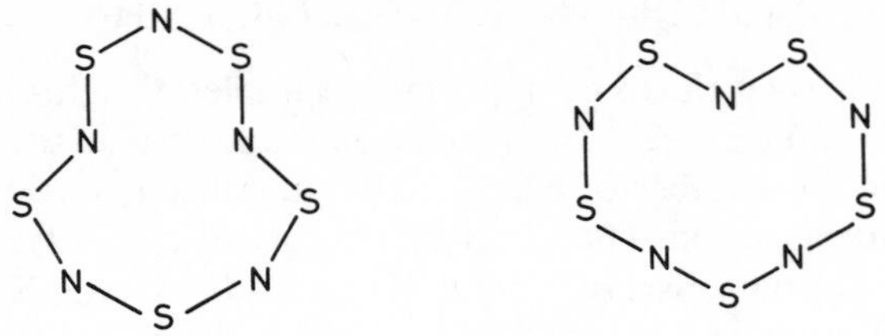

$$[S_5N_5][S_3N_3O_4] \qquad\qquad [S_5N_5][AlCl_4]$$

Figure 4.26 Conformers of the $S_5N_5^+$ cation.

As mentioned above, the cation exists in different conformations—either heart-shaped with the $AlCl_4^-$ anion, or azulene like in the case of the $S_3N_3O_4^-$ anion (Fig. 4.26). Both structures are planar; the S—N bond lengths are short, supporting the idea of aromaticity in the cation. The synthetic usefulness of salts of $S_5N_5^+$ has not yet been established.

4.5 SULPHUR–NITROGEN CAGES

4.5.1 Tetrasulphur tetranitride

The best known sulphur–nitrogen cage is tetrasulphur tetranitride, the first reports of its preparation dating back to 1835. However, it is thermodynamically unstable ($\Delta H_f = 460 \, \text{kJ mol}^{-1}$) and explodes when subjected to shock. Traces of S_4N_4 are often observed as side-products in many sulphur-nitrogen reactions. There are numerous routes for its formation and these are shown, together with routes to related cages, in Fig. 4.27. The most convenient laboratory synthesis is direct reaction of NH_3 with SCl_2/S_2Cl_2 in CCl_4.[48] This is clearly a multi-stage process and the sequence of reactions probably involves several planar SN heterocycles:

$$2S_2Cl_2 + 4NH_3 \longrightarrow NSCl + 3NH_4Cl + 3S$$

$$2NSCl + S_2Cl_2 \longrightarrow [S_3N_2Cl]Cl + SCl_2$$

$$3[S_3N_2Cl]Cl + S_2Cl_2 \longrightarrow 2[S_4N_3]Cl + 3SCl_2$$

$$[S_4N_3]Cl + 4NH_3 + 2SCl_2 \longrightarrow S_4N_4 + 3NH_4Cl + S_2Cl_2$$

The structure of S_4N_4 (Fig. 4.28) consists (in both the gas phase and the solid state) of a square plane of nitrogen atoms lying within a tetrahedron of sulphur atoms with overall D_{2d} symmetry. All of the S—N bond lengths are approximately equal (1.63 Å), and the S—S distances (2.59 and 2.71 Å), although longer

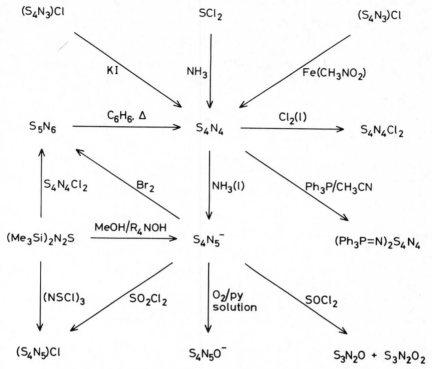

Figure 4.27 Interconversion of sulphur–nitrogen cages.

than single S—S bonds (2.05 Å), are considerably shorter than the sum of the van der Waals radii (3.70 Å).

Several detailed studies have attempted to rationalize the observed structure and also the thermochromic behaviour (at 77 K S_4N_4 is almost colourless, at 243 K canary yellow, at room temperature orange and at 373 K blood red) and the observation of a small dipole moment (0.56 D in benzene at 318 K). UV photoelectron spectroscopy provides experimental information which is very difficult to interpret and the assignments are still controversial. MO calculations of increasing complexity have been performed during the past two decades but many difficulties remain. One of the more straightforward approaches is that of Gleiter,[49] who starts by carrying out calculations for planar S_4N_4 and then considers the effect of distortion on the MOs.

Figure 4.29 shows a π–MO scheme for the hypothetical planar D_{4h} S_4N_4. As can be seen, in this conformation a triplet ground state is predicted. Distortion to the observed cradle conformation results in considerable stabilization of the a_{2u} level as a result of strong interaction of the sulphur 3p orbitals. For bond lengths

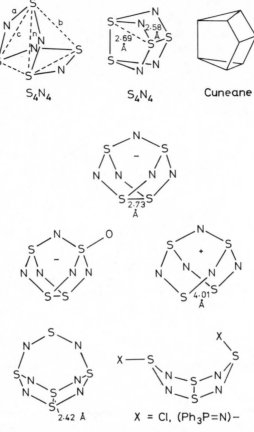

Figure 4.28 Structures of sulphur–nitrogen cages.

of between 2.00 and 2.80 Å sulphur 3p–3p overlap is more favourable than nitrogen 2p–2p overlap, leading to the observed structure with the nitrogens in a square plane and non-bonding. If we refer back to P_4S_4 and As_4S_4 (p. 60), the sulphur atoms occupy the square planar positions. Calculations show that the overlap integrals for P(3p–3p) and As(4p–4p) are larger than for S(3p–3p), and in consequence the sulphur atoms adopt the non-bonding sites in α-P_4S_4 and As_4S_4. The MO scheme in Fig. 4.29 may also be used for the 10π and 11π oxidized cations of S_4N_4 described on p. 107.

Empirical treatments can be useful. In total, S_4N_4 has 44 valence electrons, i.e. four more than required for the electron-precise eight-vertex structure of cuneane, C_8H_8. The extra two pairs of electrons go into antibonding orbitals to break two bonds in the cuneane structure.

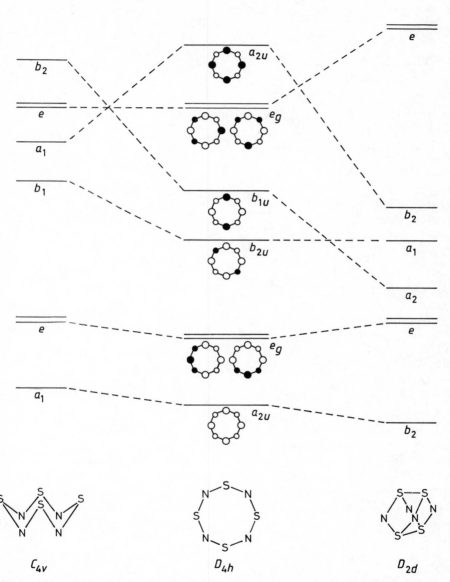

Figure 4.29 π MO schemes for the hypothetical planar, crown and known cradle structures of S_4N_4.

112

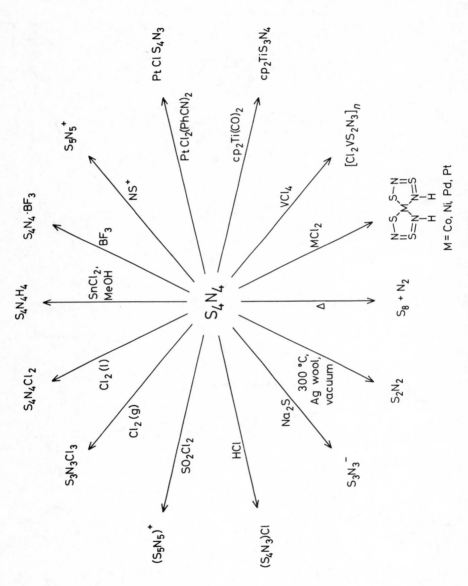

Figure 4.30 Reactions of S_4N_4.

Banister[50] has proposed a different scheme. Each atom has one lone pair and each S—N connection uses one pair of electrons, leaving six pairs of electrons, which are then placed along the edges of the S_4 tetrahedron. This approach does not properly rationalize the shortness of the S—N bonds which are considered to be 'compressed' by the bonding within the S_4 tetrahedron; additionally, all of the nitrogen atoms only have six valence electrons. Despite these limitations, it is easy to extend the treatment to related compounds. S_5N_6, $S_4N_5^+$ and $S_4N_5^-$ are related to S_4N_4 by bridging an S—S bond by an N_2S group or an N^+ or N^-, respectively. Thus in S_5N_6, of a total of 60 valence electrons 22 are non-bonding and 24 are in S—N bonds, leaving 14 electrons to place along the original S_4 tetrahedrons edges. The extra two electrons result in a shorter S—S bond (2.42 Å) as shown. Similarly, in $S_4N_5^+$ there are ten electrons available for S_4 bonding and there is one long bond (4.01 Å) whilst $S_4N_5^-$ has $12e^-$ available and hence the S—S distances are close to those in S_4N_4.

MO calculations for S_4N_4 indicate partial charges of $ca + 1$ for sulphur and $- 0.3$ for the nitrogen atoms, and therefore it is to be expected that electrophilic reagents will attack nitrogen whereas nucleophilic reactions will proceed via sulphur. In general this is the case; some reactions of S_4N_4 are shown in Fig. 4.30.

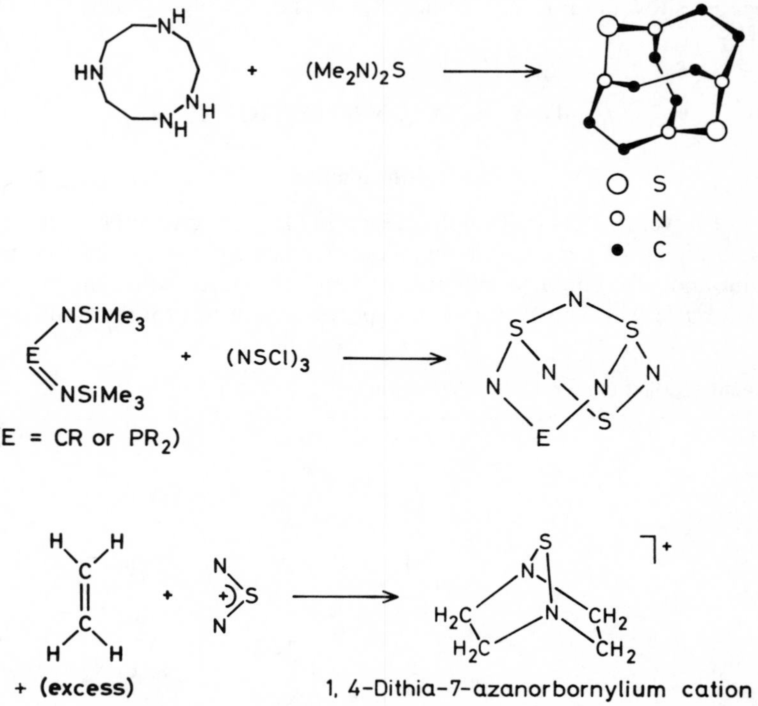

Figure 4.31 Some sulphur–nitrogen–carbon cage compounds (see ref. 51).

The cage/ring may be (1) conserved, (2) expanded, (3) contracted or (4) cleaved. Its Lewis base behaviour is via nitrogen. A wide range of adducts are known and, as might be expected from Fig. 4.29, when one of the nitrogen atoms uses two-electrons to bind a Lewis acid (e.g. BF_3) the cluster bonding count is effectively reduced, leading to a more open (crown S_8) ring structure. Thermodynamically, S_4N_4 is unstable with respect to formation of sulphur and N_2, and it can detonate when heated or subjected to shock and so must be handled with care.

Apart from the formation of other SN heterocycles, carbon-containing heterocycles referred to on p. 98 and 100 may also be prepared. There has also been substantial interest in the preparation of metalla–sulphur-nitrogen rings.[31,32] Perhaps the most obvious conclusion to be drawn is that the diversity of reactions that S_4N_4 undergoes defies rationalization. It seems likely that until new mechanistic insights are gained we are free to consider almost any reaction as a source of excitement!

4.5.2 Other Cages

Routes to $S_4N_5^+$, $S_4N_5^-$ and S_5N_6 are shown in Fig. 4.27 and the structures of the cages are shown in Fig. 4.28. Other examples of SN cages[59] are illustrated in Fig. 4.31.

4.6 POLYATOMIC CATIONS

4.6.1 Introduction

It was observed in 1804 that sulphur reacts in oleum to give intense blue, red or yellow solutions, depending on the time of reaction and the concentration. Tellurium and selenium also give rise to deeply coloured solutions in concentrated sulphuric acid. The identity of the species responsible for these colours was

Table 4.3 The known Group VI cations

M_4^{2+}	S_4^{2+}	Se_4^{2+}	Te_4^{2+}	$Te_2Se_2^{2+}$ S_3Se^{2+}	
M_6^{2+}	—	—	—	$Te_2Se_4^{2+}$ $Te_3S_3^{2+}$	$Se_6Ph_2^{2+}$ Se_6I^+
M_6^{4+}	—	—	Te_6^{4+}		$Se_6I_2^{2+}$ S_7I^+
					$(S_7I)_2I^{3+}$
M_8^{2+}	S_8^{2+}	Se_8^{2+}		$Se_6Te_2^{2+}$	
					Se_9Cl^+
M_{10}^{2+}		Se_{10}^{2+}		$Te_2Se_8^{2+}$	
M_{19}^{2+}	S_{19}^{2+}				

in doubt until recently. Gillespie and co-workers[52] have spearheaded the study of the oxidation of Group VI elements with some remarkable cations having been isolated. Table 4.3 summarizes the known species. It is clear that the overall electron count for many of the S_x^{n+} cations falls between those of (Group VI)$_x$ and S—N molecules; for example, S_8, S_8^{2+} and S_4N_4 are 48-, 46- and 44-electron species, respectively. Apart from the difficult synthetic chemistry, the various cations have presented substantial problems theoretically and the empirical treatment developed to rationalize their structures was outlined in Chapter 1 (p. 2). An excellent review of this area was given by Gillespie.[52]

4.6.2 Preparation

A number of routes are available. For success, the final anion has to be insufficiently basic to react with the cation. The cations are usually extremely sensitive to nucleophilic attack and so solvents such as SO_2, AsF_3 and HSO_3F and counter ions such as $AlCl_4^-$, AsF_6^- and SO_3F^- are essential. Typical oxidations are shown below.

$$S_8 + 3AsF_5 \xrightarrow{SO_2} [S_8]^{2+}[AsF_6]_2 + AsF_3$$

$$4Se + S_2O_6F_2 \longrightarrow Se_4^{2+} + 2SO_3F^-$$

$$SeCl_4 + 15Se + 4AlCl_3 \xrightarrow{melt} 2[Se_8][AlCl_4]_2$$

Although the above methods are logical, it is worth noting the preparation of S_{19}^{2+}:

$$28S + 5AsF_5 \longrightarrow [S_{19}^{2+}][AsF_6]_2$$

Mixed cations can also be prepared by the use of alloys, reaction of an element with a cation of a different element or reaction between cations of two different elements, although, as will be seen below, it is not all smooth sailing!

$$Se–Te(alloy) \xrightarrow{SbF_5} [Te_2Se_2^{2+}][SbF_6][Sb_3F_{14}]$$

$$Se_8^{2+} + 2Te \xrightarrow{SO_2} Se_8Te_2^{2+}$$

$$Te_4^{2+} + Se_8^{2+} \longrightarrow Te_3Se_3^{2+}$$

Halogen-substituted monocations are obtained if the oxidation is carried out in the presence of the halogen[53] and by several other routes, e.g. reaction of excess of sulphur with $I_2Sb_2F_{11}$ or by reaction of I_2 with a previously prepared cation,

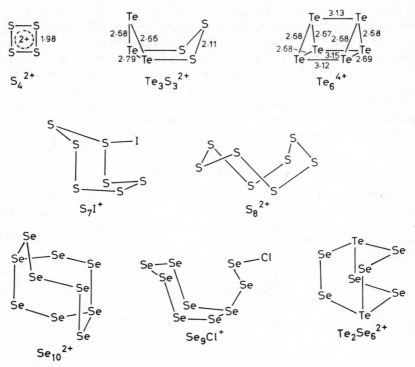

Figure 4.32 Structures of Group VI cations. Bond lengths in ångstroms.

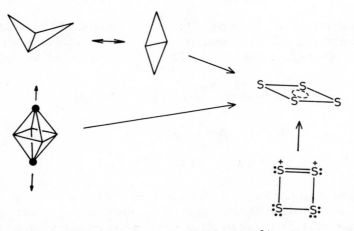

Figure 4.33 Structure and bonding in the S_4^{2+} cation.

$S_x(AsF_6)_2$ $(x \approx 19)$:[54]

$$12Se + I_2 + 3AsF_5 \xrightarrow{SO_2} 2[Se_6I][AsF_6] + AsF_3$$

$$4S_8 + 2I_2 + 9AsF_5 \longrightarrow [(S_7I)_4][S_4][AsF_6]_6 + 3AsF_3$$

In a complex reaction, grey selenium reacts with $[NO][SbCl_6]$ in liquid SO_2 to give $[Se_9Cl][SbCl_6]$.[55]

4.6.3 Structure and Bonding

Selected solid-state structures are shown in Fig. 4.32. The M_4^{2+} cations have a total electron count of $22e^-$ or 11 pairs. Using Gillespie's treatment (p. 6), this cation has $5n + 2$ electrons and should be based on a tetrahedron with one bond broken, i.e. a butterfly structure. Subtracting one non-bonding lone pair for each atom leaves seven pairs of electrons for 'cluster' bonding. Using Wade's rules we would predict an *arachno* structure similar to B_4H_{10}. The observed structure is square planar. We can post-rationalize the observed structure in both treatments. Firstly, we could have two valence forms of the butterfly structure which when combined give a planar structure. Equally, a planar cation can be derived from an octahedron by removal of two opposite vertices and may be described as *iso-arachno* (Fig. 4.33). Obviously, we should not be epecially pleased with these types of arguments—for empirical rules to be worthwhile they should predict structures without too many 'adaptations.' However, some support for a planar structure with extra electron density in the ring are the slightly short M—M bonds in M_4^{2+} (e.g. in S_4^{2+} S—S = 1.98 Å versus 2.04 Å in S_8).

The $Te_3S_3^{2+}$ cation contains 34 electrons $(5n + 4)$ and is based on a trigonal prism with two bonds broken or on P_4S_3 with one vertex removed (p. 61). The Te_6^{4+} cation has, surprisingly, a symmetrical structure. This can be rationalized as a trigonal prism with any one of three bonds being broken (or all three broken to an equal extent). Figure 4.34 illustrates the relationship of M_6^{n+} structures; notice that more electrons result in more open geometries. The structure of S_7I^+ is analogous to that of S_7O (p. 44). Two forms of M_8^{2+} $(5n + 6e^-)$ cations have now been observed. $Te_2Se_6^{2+}$ is based on a cube with three bonds broken, whilst S_8^{2+} is related to cuneane with three bonds broken (or S_4N_4 with one bond broken; cf. p. 108). The structure[56] of Se_{10}^{2+} is derived by bridging two edges of a pentagonal prism and subsequent removal of two vertices (Fig. 4.35). Similar arguments may be used to rationalize the structure[57] of S_{19}^{2+} and the interested reader is directed to an excellent review.[54]

The solution structures of Group VI cations have been probed by ^{77}Se and ^{125}Te NMR.[58-60] Both Se_8^{2+} and $Te_2Se_6^{2+}$ retain their (different) solid-state conformations in solution. However, the Se_{10}^{2+} cation does appear by ^{77}Se NMR to undergo structural isomerism with two resonances whose relative intensities are temperature dependent.

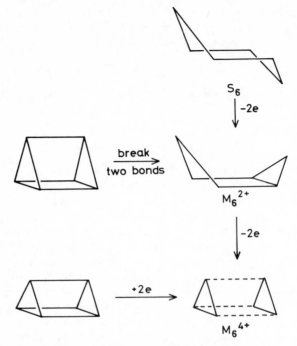

Figure 4.34 Structure of $S_6{}^{n+}$ cations.

Figure 4.35 Structure of the $Se_{10}{}^{2+}$ cation.

4.6.4 Reactions

The reaction of $[Se_4][AsF_6]_2$, $[Se_8][AsF_6]_2$ or $[Se_{10}][AsF_6]_2$ with $M(CO)_6$ (M = Mo or W) gives $[M_2(CO)_{10}Se_4][AsF_6]_2$ in which two $[M(CO)_5Se_2]^+$ cations are linked by long $Se\cdots Se$ bonds (3.015 Å)[61]. The $Se_4{}^{2+}$ cation also provides a source of an Se_2 group in the synthesis of $[FeW(CO)_8Se_2][SbF_6]_2$.[62] Reaction of $Se_4{}^{2+}$ with PhSeSePh results in a boat conformation $Se_6Ph_2{}^{2+}$

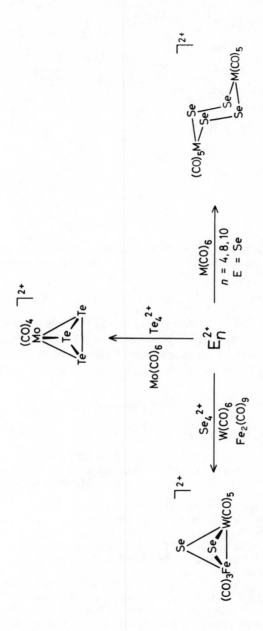

Figure 4.36 Reactions of Se_n^{2+} cations.

ring,[63] as shown below.

Finally, the usefulness of Group VI cations in organic synthesis has been described, e.g. the preparation of perfluorinated sulphides, selenides and tellurides from C_2F_4:[64]

$$[Se_8][AsF_6]_2 + C_2F_4 \rightarrow (C_2F_5)_2Se_x \; (x = 2, 3)$$

More recently,[65] the oxidizing power of S_n^{2+} cations has been demonstrated. Reaction of S_8^{2+} with hydrocarbons proceeds with insertion of sulphur into a C—H bond (e.g. CH_4 gives methanethiol); toluene is converted into a series of isomeric dimethylbiphenyls, whilst reaction with CO gives COS.

4.7 REFERENCES

1. I. Haiduc, *The Chemistry of Inorganic Ring Systems, Part 1*, Wiley–Interscience, New York, 1970.
2. (a) P. Paetzold, C. Plotho, G. Schmid and R. Buese, *Z. Naturforsch., Teil B*, 1984, **39**, 1069. (b) T. Casparis-Ebeling and H. Noth, *Angew. Chem., Int. Ed. Engl.*, 1984, **23**, 303.
3. C. Habben, A. Meller, M. Noltemeyer and G. M. Sheldrick, *J. Organomet. Chem.*, 1985, **288**, 1.
4. G. Sussfink, *Chem. Ber.*, 1986, **119**, 2393.
5. A. W. Harsberger, G. Lee, R. F. Porter and S. H. Bauer, *Inorg. Chem.*, 1969, **8**, 1683.
6. H. Bock and W. Fuss, *Angew. Chem., Int. Ed. Engl.*, 1971, **10**, 182.
7. E. K. Mellon, B. M. Coker and P. B. Dillon, *Inorg. Chem.*, 1972, **11**, 852.
8. C. K. Narula, J. F. Janik, E. N. Duesler, R. T. Paine and R. Schaeffer, *Inorg. Chem.*, 1986, **25**, 3346.
9. J. C. Sheldon and B. C. Smith, *Q. Rev. Chem. Soc.*, 1960, **14**, 200.
10. H. Stenberg and R. J. Brotherton, Boron–nitrogen and phosphorus–nitrogen compounds in *Organoboron Compounds*, Vol. 2, Wiley, London, 1967.
11. M. Scotti, M. Valderrama, R. Ganz and H. Werner, *J. Organomet. Chem.*, 1985, **286**, 399.

12. H. R. Allcock *Phosphorus–Nitrogen Compounds*, Academic Press, New York, 1972.
13. H. G. Heal, *The Inorganic Heterocyclic Chemistry of Sulfur, Nitrogen and Phosporus*, Academic Press, London, 1980.
14. R. Keat, *Top. Curr. Chem*, 1982, **102**, 89.
15. S. S. Krishamurthy and A. C. Sau, *Adv. Inorg. Radiochem.*, 1978, **21**, 41.
16. K. V. Katti, U. Seseke and H. W. Roesky, *Inorg. Chem.*, 1987, **26**, 814.
17. M. J. S. Dewar, E. A. C. Lucken and M. A. Whitehead, *J. Chem. Soc.*, 1960, 2423.
18. G. Trinquier, *J. Am. Chem. Soc.*, 1986, **108**, 568.
19. H. R. Allocock, J. L. Desorcie and G. H. Riding, *Polyhedron*, 1987, **6**, 119.
20. H. R. Allcock, M. L. Levin and P. E. Austin, *Inorg. Chem.*, 1986, **25**, 2281.
21. (a) P. Cassoux, O. Glemser and J. F. Labarre, *Z. Naturforsch., Teil B*, 1977, **32**, 41.
    (b) N. Burford, T. Chivers, P. W. Codding and R. T. Oakley, *Inorg. Chem.*, 1982, **21**, 982.
22. *Gmelin Handbook of Inorganic Chemistry, Sulfur–Nitrogen Compounds, Parts 2 and 3*, 8th ed., Springer Verlag, Heidelberg, 1984 and 1987.
23. W. L. Jolly and K. D. Maguire, *Inorg. Synth.*, 1967, **9**, 102.
24. T. Moeller, T. H. Chang, A. Ouchi, A. Vandi and A. Failli, *Inorg. Synth.*, 1972, **13**, 9.
25. J. C. Van Grampel, *Rev. Inorg. Chem.*, 1981, **3**, 1.
26. H. Oberhammer, A. Waterfeld and R. Mews, *Inorg. Chem.*, 1984, **23**, 415.
27. N. Burford, T. Chivers, M. N. S. Rao and J. F. Richardson, *Rings, Clusters and Polymers of the Main Group Elements*, ACS Symposium Series, No. 232, American Chemical Society, Washington, DC, 1983, 81.
28. M. Herberhold and L. Haumaier, *Z. Naturforsch., Teil B*, 1980, **35**, 1277.
29. A. Apblett, A. J. Banister, D. Biron, A. G. Kendrick J. Passmore, M. Schriver and M. Stojanac, *Inorg. Chem.*, 1986, **25**, 4451.
30. R. Jones, I. P. Parkin, D. J. Williams and J. D. Woollins, *Polyhedron*, in press.
31. P. F. Kelly and J. D. Woollins, *Polyhedron*, 1986, **5**, 607.
32. T. Chivers and F. Edelmann, *Polyhedron*, 1986, **5**, 1661.
33. U. Demant and K. Dehnicke, *Z. Naturforsch., Teil B*, 1986, **41**, 929.
34. *Gmelin Handbook of Inorganic Chemistry, Sulphur–Nitrogen Compounds, Part 1*, 8th ed., Springer Verlag, Heidelberg, 1977.
35. A. J. Banister, *Nature Phys. Sci.*, 1972, **237**, 92.
36. J. C. Morris and C. W. Rees, *Chem. Soc. Rev.*, 1986, 1; *Gmelin Handbook of Inorganic Chemistry, Sulphur–Nitrogen Compounds, Part 4*, 8th ed., Springer Verlag, Heidelberg, 1987.
37. T. Chivers, *Chem. Rev.*, 1985, **85**, 341; H. W. Roesky, *Angew. Chem., Int. Ed. Engl.*, 1979, **18**, 91; R. Gleiter, *Angew. Chem., Int. Ed. Engl.*, 1981, **20**, 444.
38. M. M. Labes, P. Lowe and L. F. Nichols, *Chem. Rev.*, 1979, **79**, 1.
39. E. G. Awere, N. Burford, C. Mailer, J. Passmore, M. J. Schriver, P. S. White, A. J. Banister, H. Oberhammer and L. H. Sutcliffe, *J. Chem. Soc., Chem. Commun.*, 1987, 66; W. V. Brooks, N. Burford, J. Passmore, M. J. Schriver and L. H. Sutcliffe, *J. Chem. Soc., Chem. Commun.*, 1987, 69.
40. J. Bojes, T. Chivers, W. G. Laidlaw and M. Trsic, *J. Am. Chem. Soc.*, 1979, **101**, 4517.
41. T. Chivers and R. T. Oakley, *Top. Curr. Chem.*, 1982, **102**, 117.
42. P. Dubois, J. P. Lelieur and G. Lepoutre, *Inorg. Chem.*, 1987, **26**, 1897.
43. R. W. H. Small, A. J. Banister and Z. V. Hauptman, *J. Chem. Soc., Dalton Trans.*, 1981, 2188.
44. J. Bojes, T. Chivers, A. W. Cordes, G. Maclean and R. T. Oakley, *Inorg. Chem.*, 1981, **20**, 16.
45. T. Chivers, A. W. Cordes, R. T. Oakley and P. N. Swepston, *Inorg. Chem.*, 1981, **20**, 2376.

46. I. Ernest, W. Holick, G. Rihs, D. Schomburg, G. Shohan, D. Wenkert and R. B. Woodward, *J. Am. Chem. Soc.*, 1981, **103**, 1540.
47. A. J. Banister, Z. V. Hauptman, A. G. Kendrick and R. W. H. Small, *J. Chem. Soc., Dalton Trans.*, 1987, 915, and references cited therein.
48. M. Villana-Blanco and W. L. Jolly, *Inorg. Synth.*, 1967, **9**, 98.
49. R. Gleiter, *Angew. Chem., Int. Ed. Engl.*, 1981, **20**, 445.
50. A. J. Banister, *Nature Phys. Sci.*, 1972, **239**, 69.
51. M. Magerstadt, R. B. King, M. G. Newton, N. E. Tonks and C. E. Ringold, *J. Am. Chem. Soc.*, 1986, **108**, 851; N. Burford, J. P. Johnson, J. Passmore, M. J. Schriver and P. S. Winter, *J. Chem. Soc., Chem. Commun.*, 1986, 967; R. T. Boere, R. T. Oakley and M. Shevalier, *J. Chem. Soc., Chem. Commun.*, 1987, 110.
52. R. J. Gillespie, *Chem. Soc. Rev.*, 1979, **8**, 315.
53. W. A. S. Nandana, J. Passmore and P. S. White, *J. Chem. Soc., Chem. Commun.*, 1983, 526.
54. J. Passmore, G. Sutherland, P. Taylor, T. K. Whidden and P. S. White, *Inorg. Chem.*, 1981, **20**, 3839.
55. R. Faggiani, R. J. Gillespie, J. W. Kolls and K. C. Malhotra, *J. Chem. Soc., Chem. Commun.*, 1987, 591.
56. R. C. Burns, W. L. Chan, R. J. Gillespie, W. C. Luk, J. F. Sawyer and D. R. Slim, *Inorg. Chem.*, 1980, **19**, 142.
57. R. C. Burns, R. J. Gillespie and J. F. Sawyer, *Inorg. Chem.*, 1980, **19**, 142.
58. R. C. Burns, M. J. Collins, R. J. Gillespie and G. J. Schrobilgen, *Inorg. Chem.*, 1986, **25**, 4465.
59. M. J. Collins and R. J. Gillespie, *Inorg. Chem.*, 1984, **23**, 1975.
60. M. J. Collins, R. J. Gillespie, J. F. Sawyer and G. J. Schrobilgen, *Inorg. Chem.*, 1986, **25**, 2053.
61. M. J. Collins, R. J. Gillespie, J. W. Kolis and J. F. Sawyer, *Inorg. Chem.*, 1986, **25**, 2057.
62. D. J. Jones, T. Makani and J. Roziere, *J. Chem. Soc., Chem. Commun.*, 1986, 1275.
63. R. Faggiani, R. J. Gillespie and J. W. Kolsis, *J. Chem. Soc., Chem. Commun.*, 1987, 592.
64. C. D. Desjardins and J. Passmore, *J. Chem. Soc., Dalton Trans.*, 1973, 2314.
65. A. M. Rosan, *J. Chem. Soc., Chem. Commun.*, 1985, 377.

See also '*The Chemistry of Inorganic Homo- and Heterocycles*', Vols. I and II, 1987, I. Haiduc and D. B. Sowerby, Academic Press, London.

Index